Lecture Notes in Computer Science 14812

Founding Editors

Gerhard Goos
Juris Hartmanis

The series Lecture Notes in Computer Science (LNCS), including its subseries Lecture Notes in Artificial Intelligence (LNAI) and Lecture Notes in Bioinformatics (LNBI), has established itself as a medium for the publication of new developments in computer science and information technology research, teaching, and education.

LNCS enjoys close cooperation with the computer science R & D community, the series counts many renowned academics among its volume editors and paper authors, and collaborates with prestigious societies. Its mission is to serve this international community by providing an invaluable service, mainly focused on the publication of conference and workshop proceedings and postproceedings. LNCS commenced publication in 1973.

Hao Chen · Yuyin Zhou · Daguang Xu ·
Varut Vince Vardhanabhuti
Editors

Trustworthy Artificial Intelligence for Healthcare

Second International Workshop, TAI4H 2024
Jeju, South Korea, August 4, 2024
Proceedings

 Springer

Editors
Hao Chen (iD)
The Hong Kong University of Science
and Technology
Kowloon, Hong Kong

Daguang Xu (iD)
NVIDIA Corporation
Santa Clara, CA, USA

Yuyin Zhou (iD)
University of California, Santa Cruz
Santa Cruz, CA, USA

Varut Vince Vardhanabhuti
University of Hong Kong
Pok Fu Lam, Hong Kong

ISSN 0302-9743   ISSN 1611-3349 (electronic)
Lecture Notes in Computer Science
ISBN 978-3-031-67750-2   ISBN 978-3-031-67751-9 (eBook)
https://doi.org/10.1007/978-3-031-67751-9

This Springer imprint is published by the registered company Springer Nature Switzerland AG
The registered company address is: Gewerbestrasse 11, 6330 Cham, Switzerland

If disposing of this product, please recycle the paper.

Preface

We are pleased to present the proceedings of the Second International Workshop on Trustworthy Artificial Intelligence for Healthcare (TAI4H), which was held in conjunction with the 33rd International Joint Conference on Artificial Intelligence (IJCAI) on August 4, 2024.

Artificial intelligence (AI) has achieved or even exceeded human performance in many healthcare tasks, owing to the fast development of AI techniques and the growing scale of medical data. However, AI techniques are still far from being widely applied in healthcare practice. Real-world scenarios are far more complex, and AI is often faced with challenges in its credibility such as lack of explainability, generalization, fairness, privacy, etc. The development of trustworthy artificial intelligence for healthcare (TAI4H) is hence of great importance to enhance the trust and confidence of doctors and patients in using the related techniques. We aim to bring together researchers from interdisciplinary fields, including but not limited to machine learning, clinical research, and medical imaging, etc., to provide different perspectives on how to develop trustworthy AI algorithms to accelerate the adoption of AI in healthcare.

IJCAI 2024 TAI4H attracted 21 valid submissions. These submissions were selected through a double-blind peer-review process. We adopted the Conference Management Toolkit (CMT) for paper submissions and double-blind peer review.

IJCAI 2024 TAI4H had 4 Program Committee Chairs, an 8-person Program Committee, a 6-person Advisory Committee, and 2 Student Organizers. 10 Area Chairs (AC) were responsible for organizing paper reviews. Each area chair was assigned about 2 manuscripts, for each of which they were asked to suggest up to 2–3 potential reviewers. Subsequently, over 22 invited reviewers were invited. Final reviewer allocations via CMT took account of PC suggestions and reviewer bidding, finally allocating about 2 papers per reviewer. Based on the double-blinded reviews, Area Chairs' recommendations, and Program Chairs' overall adjustments, 5 papers (24%) were accepted as oral presentations, 8 papers (38%) were accepted as poster presentations, and 8 papers (38%) were rejected. This process resulted in the acceptance of a total of 13 papers, reaching an overall acceptance rate of 62% for IJCAI 2024 TAI4H.

We would like to express our gratitude to all the authors who submitted and presented their outstanding work at TAI4H, which made TAI4H a resounding success. We appreciate the professional and constructive review comments from all the reviewers. We were also honoured to have invited distinguished speakers to deliver their insightful talks. We hope these high-quality works and in-depth discussions will pave new pathways and inspire new directions for the trustworthy AI for healthcare.

August 2024

Hao Chen
Yuyin Zhou
Daguang Xu
Varut Vince Vardhanabhuti

Organization

Program Committee Chairs

Hao Chen	The Hong Kong University of Science and Technology, China
Yuyin Zhou	University of California, Santa Cruz, USA
Daguang Xu	NVIDIA, USA
Varut Vince Vardhanabhuti	University of Hong Kong, China

Program Committee

Luyang Luo	The Hong Kong University of Science and Technology, China
Yueming Jin	National University of Singapore, Singapore
Lequan Yu	University of Hong Kong, China
Junlin Hou	The Hong Kong University of Science and Technology, China
Jiguang Wang	The Hong Kong University of Science and Technology, China
Xi Wang	The Chinese University of Hong Kong, China
Jing Qin	The Hong Kong Polytechnic University, China
Ziyue Xu	NVIDIA, USA

Advisory Committee

Marius George Linguraru	Children's National Hospital, USA
Le Lu	Alibaba, China
Danny Z. Chen	University of Notre Dame, USA
Pheng-Ann Heng	The Chinese University of Hong Kong, China
Tim Kwang-Ting Cheng	The Hong Kong University of Science and Technology, China
Stephen T. Wong	Weill Cornell Medical College, USA

Student Organizers

Wenqiang Li	The Hong Kong University of Science and Technology, China
Yu Cai	The Hong Kong University of Science and Technology, China

Reviewers

Dwarikanath Mahapatra
Fuying Wang
Huajun Zhou
Jiayu Guo
Linh Le
Nguyen Quoc Khanh Le
Qingqiu Li
Runtian Yuan
Weiwei Tian
Xuefeng Ni
Zhiyuan Cai

Fan Xiao
Haoxuan Che
Jiacheng Wang
Jie Lian
Li Liu
Nan Zhang
Pegah Ahadian
Rongjun Ge
Sicen Liu
Xinrui Jiang
Yuting He

Contents

AI Trustworthy Challenges in Drug Discovery

Pegah Ahadian and Qiang Guan[(⊠)]

Kent State University, Kent, OH, USA
{Pahadian,Qguan}@kent.edu

Abstract. Artificial intelligence (AI), including machine learning (ML) and deep learning (DL), is revolutionizing drug discovery by enabling researchers to analyze massive datasets, predict molecular behavior, and accelerate the development of new drugs. However, integrating AI into this domain presents significant computer science challenges. These challenges center on ensuring the trustworthiness of algorithms, maintaining data integrity, achieving model transparency, and upholding ethical principles. This paper explores the crucial role of AI technologies in improving drug discovery workflows while identifying the key obstacles to guaranteeing the reliability and ethical application of these technologies.

Keywords: Drug Discovery · Trustworthy · Artificial Intelligence

1 Introduction

The development of new drugs remains a time-consuming and resource-intensive endeavor, often plagued by high failure rates. Traditional methods rely heavily on laborious experimentation and can struggle to identify promising drug candidates from the vast ocean of potential molecules. This limited ability to effectively analyze large datasets and accurately predict drug-target interactions leads to several critical challenges, such as limited treatment options, slow development timelines, and high costs of drug development [6]. Artificial Intelligence (AI), encompassing Machine Learning (ML) and Deep Learning (DL), presents a powerful solution to these challenges. AI has the potential to analyze massive datasets, predict molecular interactions, and streamline drug development pipelines [9,27]. The integration of AI into drug discovery has the potential to transform the field. By addressing the limitations of traditional methods, AI offers the promise of faster development of more effective treatments, potentially at a lower cost. This could significantly improve patient outcomes and address the unmet medical needs that plague our healthcare systems.

The application of AI in drug discovery promises a revolution in healthcare. However, integrating AI into this complex domain presents challenges that go beyond technical hurdles. This paper delves into the ethical considerations and transparency concerns that emerge at this intersection. By proactively addressing these challenges, we can ensure the trustworthy implementation of AI in drug

H. Chen et al. (Eds.): TAI4H 2024, LNCS 14812, pp. 1–12, 2024.
https://doi.org/10.1007/978-3-031-67751-9_1

discovery. Traditional drug discovery methods have faced criticism for overlooking certain populations or diseases due to factors like limited commercial viability [12]. The concern is that AI, if not carefully implemented, could exacerbate these existing disparities. Algorithmic bias, data integrity issues, and the "black box" nature of complex models can all lead to unfair outcomes in drug development. Beyond the hypothetical scenario, there have been documented cases of bias in traditional drug development. For instance, a 2011 study published [14] found that women were significantly less likely than men to be enrolled in clinical trials for cardiovascular disease, despite being equally susceptible. This historical bias in research could be reflected in datasets used to train AI models, potentially perpetuating the underrepresentation of women in future drug development efforts.

This example highlights how historical biases in traditional drug discovery can be inadvertently incorporated into AI models if not addressed proactively. By including a real-world case, the argument becomes more compelling and emphasizes the importance of building trustworthy AI systems in this domain. By exploring the unique challenges of integrating AI into drug discovery, this paper aims to shed light on the potential for bias and the critical need for data integrity and fairness, the importance of model interpretability and transparency in building trust in AI, and the ethical and privacy considerations surrounding AI-powered drug development [17,19]. The ethical considerations surrounding AI-powered drug discovery necessitate a focus on four key aspects: fairness, robustness, privacy, and explainability. These pillars form the foundation for ensuring responsible and ethically sound AI use in this domain. By focusing on these four keys, we can build trustworthy AI systems that empower researchers to make responsible decisions throughout the drug discovery process. This ensures that advancements in AI contribute to the development of safe, effective, and equitable treatments for all.

This paper investigates these challenges through the lens of Quantitative Structure-Activity Relationship (QSAR) models, a common AI technique in drug discovery. We examine how different QSAR model algorithms can impact fairness and robustness. Furthermore, we explore the implications of privacy using a Natural Language Processing (NLP) based Named Entity Recognition (NER) model within the drug discovery domain. Finally, we leverage the findings on fairness, robustness, and the interpretability of the QSAR model to demonstrate the importance of explainability in building trust in AI for drug discovery.

2   Fundamentals

AI and Machine Learning (ML) systems in drug discovery must adhere to four key trustworthiness principles: fairness, robustness, privacy, and explainability. These principles are crucial for ethical AI technologies and their responsible use in advancing medical science. A structured evaluation framework helps assess and improve these principles, identifying potential weaknesses and aligning AI operations with ethical standards and societal values [5]. This framework is especially important in drug discovery, where decisions significantly impact patient health and well-being. Ensuring trustworthy AI in this field is not just a technical challenge but a moral imperative, promoting innovation grounded in ethical practices. We will explore each of these principles - fairness, robustness, privacy, and explainability - in the following sections.

Fairness. Fairness ensures AI and ML models make unbiased decisions without perpetuating inequalities based on race, gender, age, or other characteristics [2,4]. This involves ensuring that models do not favor specific populations and that their decisions are unbiased across diverse demographics. For example, drug discovery models must account for diverse genetic data to be effective across different groups, avoiding biases that could disadvantage underrepresented populations.

Robustness. Robustness refers to an AI system's ability to maintain performance across varied conditions and handle data variations or adversarial inputs without significant degradation. In drug discovery, robustness is critical as models must reliably predict drug efficacy and safety despite variations in data quality and biological variability.

Privacy. Privacy in AI focuses on protecting sensitive data, such as genetic information and medical records, ensuring models do not expose private details. Techniques like differential privacy help safeguard personal information. In drug discovery, protecting privacy is crucial to prevent the reverse engineering of data used to train models, thereby securing individual patient information [21,26].

Explainability. Explainability involves understanding and articulating the mechanisms behind AI and ML model decisions. It fosters transparency and trust by enabling users to comprehend how outputs are generated, crucial for aligning AI systems with human values and ethical standards. In drug discovery, explainability is vital for validating model predictions, identifying potential drug targets, and understanding side effects, which also supports regulatory approval by providing clear rationales for decisions.

3   Methodology

As AI plays an increasingly prominent role in drug discovery, ensuring its trustworthiness becomes critical. Addressing challenges related to fairness, robustness, privacy, and explainability is paramount. Focusing on these principles strengthens trust in AI applications, fosters ethical and reliable advancements, ensures applicability across diverse populations, safeguards sensitive data, and delivers interpretable insights that guide drug development. This study proposes utilizing Quantitative Structure-Activity Relationship (QSAR) models, a cornerstone of computational drug discovery, to assess both fairness and robustness. These models, representing the fusion of AI and Machine Learning (ML), predict the effects of chemical compounds by analyzing their structures. The underlying principle is that a compound's biological activity is quantitatively linked to its chemical structure. By leveraging QSAR methodology, we can establish relationships between a compound's structural properties (descriptors) and its biological activity, a common approach in drug discovery [11]. To complement our analysis and address the privacy metric, we incorporate Named Entity Recognition (NER) techniques [8]. NER identifies and extracts sensitive information from textual data, helping safeguard privacy concerns within drug discovery research. In addition to fairness, robustness, and privacy, explainability is a crucial metric for assessing AI trustworthiness in drug discovery. Explainability focuses on elucidating the decision-making process behind predictions, enhancing transparency and interpretability. We address this metric by leveraging the same QSAR method used for Fairness and Robustness. This allows a comprehensive examination of how the model arrives at its conclusions, significantly contributing to its ethical integrity and reliability.

4   Experimental Result on Four Dimensions

In this section, we show the experimental results obtained from evaluating our models through the lens of the four key trustworthiness principles: fairness, robustness, privacy, and explainability. Each subsequent section will focus on a specific principle, analyzing the performance of our models and exploring potential shortcomings or biases. By dissecting these results, we aim to gain a comprehensive understanding of the trustworthiness of our AI-driven approach in drug discovery.

4.1   Fairness

This section explores fairness considerations in a QSAR model for fish toxicity prediction using logistic regression [20]. We assessed fairness by comparing model performance metrics (accuracy, precision, recall, FPR), Table 1, across different groups within the dataset. The fish toxicity dataset comprises 909 chemical compounds with features including CIC0, SM1_Dz(Z), GATS1i, NdsCH, NdssC, MLOGP, and the toxicological response LC50, expressed as -LOG(mol/L). This data is used to train QSAR models that predict the environmental toxicity of chemicals based on their molecular properties. Preprocessing included normalization and handling of missing values to ensure robust model training and testing. Initially, we divided the data based on a median split of a specific feature ("CIC0"). While overall accuracy was good, we observed disparities between groups. Group 0 exhibited better accuracy and precision, suggesting a cautious approach to predictions. Conversely, Group 1 had higher recall but a significantly higher FPR, indicating a higher chance of misclassifying non-toxic compounds as toxic. This pattern suggests potential bias favoring Group 0, Table 1 summarizes the results.

Further investigation using LC50 values (lethal concentration for 50% mortality) revealed similar disparities when dividing the data based on median MLOGP (a measure of hydrophobicity). Group 1 (higher MLOGP) showed higher recall and FPR, suggesting better identification (but also misclassification) of toxic compounds. These findings highlight the importance of fairness evaluation in

QSAR models, Table 2 presents the results for Groups 0 (lower MLOGP) and 1 (higher MLOGP). Observed performance variations indicate areas for improvement, such as revisiting the modeling approach, incorporating fairness into training, and continuously monitoring performance across subgroups, Fig. 1.

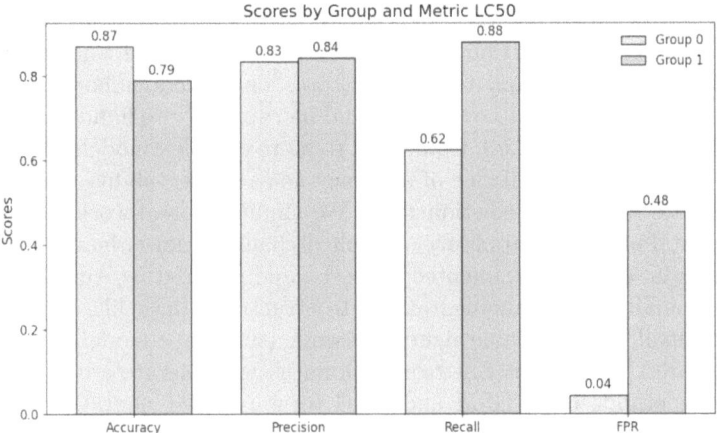

Fig. 1. Fairness Examine: LC50 Comparison Result of Two Groups

Table 1. Fairness Metrics for Fish Toxicity Prediction/CIC0

Group	Accuracy	Precision	Recall	FPR
0	0.8571	0.9677	0.7143	0.0204
1	0.7912	0.8800	0.7719	0.1765

Table 2. Fairness Metrics for Fish Toxicity Prediction/ LC50

Group	Accuracy	Precision	Recall	FPR
0	0.8696	0.8333	0.6250	0.0441
1	0.7889	0.8429	0.8806	0.4783

These findings highlight the importance of fairness evaluation in QSAR models for drug discovery. The observed disparities in performance metrics across groups indicate areas where fairness can be improved. Future work may involve revisiting the modeling approach, incorporating fairness considerations into training, and continuously monitoring model performance across different subgroups.

4.2   Robustness

A robust QSAR model for fish toxicity prediction should maintain its accuracy even under challenging conditions, such as variations in real-world data (distribution shifts) or intentional manipulations (adversarial attacks). While traditional robustness evaluation methods are designed for differentiable models like neural networks, our Random Forest model requires creative adaptation. We addressed this challenge by Distribution Shift and Augmentation and Surrogate Models for Adversarial Attacks and CLEVER Score. The Random Forest model was selected for its robustness to overfitting and its capability to handle the dataset's non-linearities effectively [24]. Compared to more complex models like deep neural networks, it offers a balance of accuracy and interpretability, crucial for the environmental toxicity prediction task. We simulated real-world variations or environmental effects by applying controlled changes (noise, feature scaling) to the input data, creating augmented datasets, and inspired by Augmix for image data, we combine these augmentations to structured data like ours. Also, we trained a small, differentiable neural network (surrogate model) to mimic the Random Forest's behavior. Surrogate allows us to Generate adversarial examples using methods like FGSM and PGD (originally for neural networks) [28] to test the model's robustness against such attacks, and estimate the CLEVER score, a metric for robustness to adversarial examples, for the non-differentiable Random Forest model.

We measured the change in prediction error as we introduced minor variations to the input data (sensitivity analysis) to assess the model's stability and sensitivity to small perturbations that providing insights into robustness. The Random Forest model exhibits higher sensitivity values compared to the surrogate model across all features. This suggests the Random Forest might be more susceptible to small variations in the input data. The lower sensitivity of the surrogate model is promising for robustness against small data perturbations, Fig. 2. However, it doesn't necessarily translate to superior overall performance. It simply indicates a different response to input changes. For the Random Forest, the third feature has the highest sensitivity, suggesting it's a crucial predictor of fish toxicity in this dataset. Evaluating robustness in non-differentiable models requires creative adaptation. By employing surrogate models, sensitivity analysis, and data augmentation, we gained valuable insights into the robustness of our Random Forest-based QSAR model. This knowledge helps enhance its reliability and trustworthiness for real-world fish toxicity prediction. AI based models in drug discovery face robustness challenges. They can be overly sensitive to slight variations in molecular structures or experimental noise, leading to unreliable predictions for new drug candidates. Additionally, these models might be vulnerable to manipulation or struggle to perform well with data from different sources.

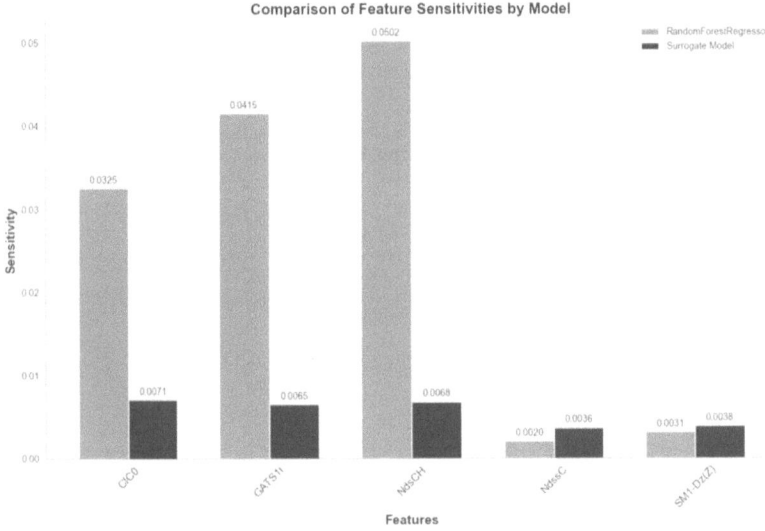

Fig. 2. Robustness Examine: Sensitivity Comparison Result of Two Models

4.3 Privacy

Human genome data offers a treasure trove of information that can be incredibly valuable in the field of drug discovery [10]. This data provides insights into the genetic basis of diseases, individual responses to medications, and pathways involved in health and disease [1]. Medical documents, like Human Genome [10], contain sensitive information linked to patients. Privacy concerns the safeguarding of this sensitive data that is utilized throughout the research and development processes [15]. Data de-identification is a technique that alters healthcare data by breaking links between data and the individual it is associated with [7,16]. It removes personal information from datasets (Medical records, reports, or media) containing patients' Protected Health Information (PHI). Medical Data De-identification allows data sharing for secondary uses research and analysis while safeguarding patient privacy [16]. While Natural Language Processing (NLP) offers valuable tools like Named Entity Recognition (NER) [8] for anonymizing patient data, it has limitations. NER systems may miss crucial information [18], like Incomplete Training Data, Contextual Errors, Adversarial Attacks, and Data Re-identification. In this section we examine Named Entity Recognition (NER) [18], which entities can include names of people, organizations, locations, dates, and many other specifics. This section explores Named Entity Recognition (NER) for anonymizing patient data in drug discovery. However, NER has limitations. Figures 3 and 4 illustrate how NER systems might miss crucial details like genetic markers (Fig. 3) or patient contact information (Fig. 4). These shortcomings highlight the need for improved techniques to ensure comprehensive de-identification of sensitive information in AI-driven drug discovery.

Anonymized Text:

Patient [REDACTED], ID [REDACTED], has several notable genetic variants including rs1799971 associated with opioid addiction, and a mutation i
n the [REDACTED] gene linked to breast cancer. His genomic profile is detailed in dataset GDS000123 stored at GenBank.

Fig. 3. Output of the NER system showing Incomplete De-identification of personal identifiers, Genetic ID and Storage.

Anonymized Text:

Dr. [REDACTED] reviewed the case of patient [REDACTED] on [REDACTED], discussing his diagnosis related to an uncommon genetic marker, RYR1, wh
ich is often linked to malignant hyperthermia. Contact details for [REDACTED] are noted as alan.t@example.com and 555-0102.

Fig. 4. Output of the NER system showing Incomplete De-identification of personal identifiers, contact details.

While Named Entity Recognition (NER) has proven to be a valuable tool for the de-identification of medical data, its limitations are showed by the challenges of Incomplete De-identification. These instances reflect the necessity for enhanced training datasets, more sophisticated context-aware algorithms, and a combination of AI with rule-based approaches to ensure the comprehensive anonymization of sensitive information.

4.4 Explainability

Explainability plays a critical role in building trust in AI models used for drug discovery, particularly when considering fairness and robustness [22]. Building on the robustness and fairness assessments detailed in Sections Fairness and Robustness, this section explores the challenges of explainability in our QSAR model used for predicting fish toxicity. The results from these sections not only demonstrate the model's performance but also highlight discrepancies that need clear explanations to be fully understood and accepted by stakeholders [13]. The QSAR model itself might be a complex machine learning model, which can be opaque and difficult to interpret. This "black box" nature makes it challenging to understand the complete decision-making process behind the model's predictions [25]. The QSAR model's performance on the fish toxicity prediction task revealed potential fairness concerns. The model exhibited different accuracy, precision, recall, and FPR values for groups defined by features like CIC0 and MLOGP. Explainability techniques, such as feature importance analysis, can be employed to delve deeper into these disparities. By analyzing which features significantly influence the model's predictions for each group, researchers can pinpoint potential biases. For instance, high feature importance for CIC0 in one group compared to another might suggest the model is unfairly penalizing compounds with certain CIC0 values, leading to lower recall (missing actual toxic compounds). Similarly, high feature importance for MLOGP could indicate a bias towards over-predicting toxicity for more hydrophobic compounds (high MLOGP group with high FPR).

The SHAP summary plot (Fig. 5) highlights that MLOGP, CIC0, and SM1_Dz(Z) are the most influential features in predicting fish toxicity. It reveals that higher values of MLOGP and CIC0 tend to increase the predicted toxicity, suggesting these features play a critical role in the model's decision-making process. Counterfactual explanations can be particularly valuable in addressing fairness concerns [23]. In the QSAR model example, these techniques could be used to understand how small changes in a compound's CIC0 or MLOGP values might affect the model's prediction of fish toxicity. If a slight decrease in CIC0 significantly increases the model's predicted toxicity for a specific compound in Group 0 (typically penalized by the model), it suggests the model might be overly sensitive to this feature in that group. Conversely, if a minor increase in MLOGP doesn't significantly alter the predicted toxicity for a compound in Group 1 (prone to over-prediction), it highlights a potential bias towards under-estimating toxicity for this group.

The QSAR model evaluation also emphasized the importance of robustness [24]. Techniques like sensitivity analysis can be used to assess how the model's predictions for fish toxicity change when presented with data points containing small variations in features like CIC0 or MLOGP. If the model's predictions become unreliable or drastically different due to these minor changes, it suggests a lack of robustness. This information can guide efforts to improve the model's ability to handle real-world data variations, ultimately leading to more robust predictions. Understanding how features like CIC0 and MLOGP influence the model's decisions allows for a more informed evaluation of potential candidates for further testing. This transparency fosters trust in the AI-driven approach and enables researchers to make well-founded decisions throughout the drug discovery process, ultimately contributing to the development of safe and reliable drugs. Explainability serves as a bridge between the complex AI model and human understanding. The complexity of black box nature makes it challenging to understand the complete decision-making process behind the model's predictions and researchers should identify and address potential biases, evaluate robustness to support trustworthy decision-making in drug discovery.

5 Enhancement and Future Work

Building trust in AI models for drug discovery goes beyond addressing individual challenges like fairness and explainability. It necessitates a comprehensive approach that integrates these considerations throughout the entire development process. This includes employing robust data collection practices to minimize bias from the outset. Techniques like gathering data from diverse populations and employing balancing techniques within datasets can help mitigate bias in the foundation upon which the model is built. Furthermore, ongoing monitoring and human oversight are crucial for maintaining trust in AI-driven drug discovery. Regularly evaluating the model's performance for fairness and robustness is essential. Human experts should be involved in interpreting the model's outputs and making final decisions, particularly in critical stages of drug development.

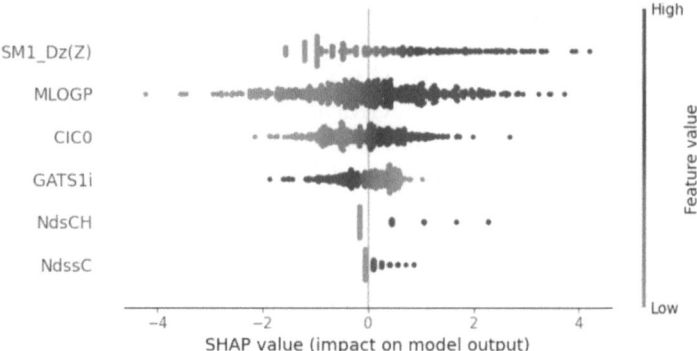

Fig. 5. SHAP Plot: Shows how individual features affect the model's predictions. The color of the points indicates the feature value (red for high, blue for low). (Color figure online)

This collaborative approach leverages the strengths of both AI and human expertise, ultimately fostering trust in the drug discovery process. Additionally, open communication and transparency throughout the AI development process are essential for building trust with stakeholders, including regulatory bodies and the public. By clearly communicating the model's limitations and potential biases, researchers can foster a collaborative environment for addressing concerns and ensuring the responsible use of AI in drug discovery. While this paper explores the individual challenges of fairness, robustness, privacy, and explainability in AI-driven drug discovery, building trust necessitates a more holistic approach. Future work will focus on developing a comprehensive framework that integrates these considerations throughout the entire development process, from data collection to model deployment.

6   Conclusion

The potential of AI to revolutionize drug discovery is undeniable, but as this paper has emphasized, ensuring the trustworthiness of these models is paramount. We addressed four key challenges to trustworthiness: fairness, robustness, privacy, and explainability. Through the lens of data analysis, model evaluation, and specific techniques like sensitivity analysis, we explored how to mitigate bias, ensure model reliability, and protect patient privacy. Additionally, we highlighted the importance of explainability techniques like feature importance analysis and counterfactual explanations in fostering trust and guiding researchers towards well-informed decisions. This paper's major contribution lies in proposing a comprehensive framework for building trustworthy AI in drug discovery. This framework emphasizes the importance of integrating fairness, robustness, privacy, and explainability considerations throughout the entire development process, from data collection to model deployment. The case study using a QSAR model for fish toxicity prediction serves as a practical example of how this framework can be applied. By demonstrating how explainability techniques can identify potential biases and evaluate robustness, the case study underscores the framework's effectiveness in building trustworthy AI for drug discovery.

This work highlights the need for further research on explainability techniques tailored to drug discovery data and their integration throughout the development pipeline. Additionally, fostering collaboration between AI researchers, drug discovery scientists, and regulatory bodies is crucial for establishing best practices and ensuring the ethical and successful implementation of trustworthy AI in drug discovery. By addressing these challenges, we can harness the full potential of AI to accelerate the development of safe and effective medications, ultimately benefiting patients worldwide.

Acknowledgement. This work is partially sponsored by NSF #2212465.

References

1. Tabakhi, S., Suvon, M.N., Ahadian, P., Lu, H.: Multimodal learning for multi-omics: a survey. World Sci. Annu. Rev. Artif. Intell. **16**(1), 2250004 (2023)
2. Ahadian, P., Parand, K.: Support vector regression for the temperature-stimulated drug release. Chaos Solitons Fractals **1**(165), 112871 (2022)
3. Walters, W.P., Barzilay, R.: Critical assessment of AI in drug discovery. Expert Opin. Drug Discov. **16**(9), 937–47 (2021)
4. Schreiner, A., Kemmerzell, N.: Towards a Quantitative Evaluation Framework for Trustworthy AI in Facial Analysis (2024)
5. Blanco-Gonzalez, A., et al.: The role of AI in drug discovery: challenges, opportunities, and strategies. Pharmaceuticals **16**(6), 891 (2023)
6. Pushpakom, S., et al.: Drug repurposing: progress, challenges and recommendations. Nat. Rev. Drug Discov. **18**(1), 41–58 (2019)
7. Prokosch, H.U., Ganslandt, T.: Perspectives for medical informatics. Methods Inf. Med. **48**(01), 38–44 (2009)
8. Song, B., Li, F., Liu, Y., Zeng, X.: Deep learning methods for biomedical named entity recognition: a survey and qualitative comparison. Briefings Bioinform. **22**(6), bbab282 (2021)
9. Qureshi, R., et al.: AI in drug discovery and its clinical relevance. Heliyon (2023)
10. Debouck, C., Metcalf, B.: The impact of genomics on drug discovery. Annu. Rev. Pharmacol. Toxicol. **40**(1), 193–208 (2000)
11. Muratov, E.N., et al.: QSAR without borders. Chem. Soc. Rev. **49**(11), 3525–64 (2020)
12. Vatansever, S., et al.: Artificial intelligence and machine learning-aided drug discovery in central nervous system diseases: state-of-the-arts and future directions. Med. Res. Rev. **41**(3), 1427–73 (2021)
13. Adadi, A., Berrada, M.: Peeking inside the black-box: a survey on explainable artificial intelligence (XAI). IEEE Access **16**(6), 52138–60 (2018)
14. Mosca, L., Barrett-Connor, E., Kass, W.N.: Sex/gender differences in cardiovascular disease prevention: what a difference a decade makes. Circulation **124**(19), 2145–54 (2011)

15. Wang, S., et al.: Genome privacy: challenges, technical approaches to mitigate risk, and ethical considerations in the United States. Ann. N. Y. Acad. Sci. **1387**(1), 73–83 (2017)
16. Tucker, K., et al.: Protecting patient privacy when sharing patient-level data from clinical trials. BMC Med. Res. Methodol. **16**, 5–14 (2016)
17. Ueda, D., et al.: Fairness of artificial intelligence in healthcare: review and recommendations. Jpn. J. Radiol. **42**(1), 3–15 (2024)
18. Catelli, R., Esposito, M.: De-identification techniques to preserve privacy in medical records. In: Artificial Intelligence in Healthcare and COVID-19, 1 January 2023, pp. 125–148. Academic Press (2023)
19. Kirby, J.B., Taliaferro, G., Zuvekas, S.H.: Explaining racial and ethnic disparities in health care. Med. Care **44**(5), 1–64 (2006)
20. Li, Y., et al.: 4D-fingerprint categorical QSAR models for skin sensitization based on the classification of local lymph node assay measures. Chem. Res. Toxicol. **20**(1), 114–28 (2007)
21. Mehrban, A., Ahadian, P.: Malware Detection in IOT Systems Using Machine Learning Techniques. arXiv preprint arXiv:2312.17683 (2023)
22. Jaganathan, K., Tayara, H., Chong, K.T.: An explainable supervised machine learning model for predicting respiratory toxicity of chemicals using optimal molecular descriptors. Pharmaceutics **14**(4), 832 (2022)
23. Jiménez-Luna, J., Grisoni, F., Schneider, G.: Drug discovery with explainable artificial intelligence. Nat. Mach. Intell. **2**(10), 573–84 (2020)
24. Polanski, J., Bak, A., Gieleciak, R., Magdziarz, T.: Modeling robust QSAR. J. Chem. Inf. Model. **46**(6), 2310–8 (2006)
25. Lundberg, S.M., Lee, S.I.: A unified approach to interpreting model predictions. In: Advances in Neural Information Processing Systems, vol. 30 (2017)
26. Bonomi, L., Huang, Y., Ohno-Machado, L.: Privacy challenges and research opportunities for genomic data sharing. Nat Genet. **52**(7), 646–654 (2020). https://doi.org/10.1038/s41588-020-0651-0. Epub 2020 Jun 29. PMID: 32601475; PMCID: PMC7761157
27. Bagheri Rajeoni, A., Pederson, B., Clair, D.G., Lessner, S.M., Valafar, H.: Automated measurement of vascular calcification in femoral endarterectomy patients using deep learning. Diagnostics **13**(21), 3363 (2023). https://doi.org/10.3390/diagnostics13213363
28. Arnab, A., Miksik, O., Torr, P.H.: On the robustness of semantic segmentation models to adversarial attacks. In: Proceedings of the IEEE Conference on Computer Vision and Pattern Recognition, pp. 888–897 (2018)

ODR3DNet: Omni-Dimension Dynamic Residual 3D Net for Pulmonary Nodule Detection

Ying Wang, Yun Tie$^{(\boxtimes)}$, Dalong Zhang, Zepeng Zhang, and Lin Qi

The School of Information and Engineering, Zhengzhou University, Zhengzhou, Henan, China
ieytie@zzu.edu.cn

Abstract. Lung cancer remains a potentially fatal global health concern, underscoring the criticality of enhancing the precision of early lung cancer diagnosis to ameliorate patient prognoses. Over the past years, the ascent of deep learning (DL) has ushered in a formidable arsenal, with DL-based computer-aided systems orchestrating remarkable strides in the realm of pulmonary nodule detection (PND). This investigation introduces a pioneering approach, namely the Omni-dimension Dynamic Residual 3D Net (ODR3DNet), meticulously tailored for PND by harnessing the prowess of full-dimensional dynamic 3D convolution. The core thrust behind ODR3DNet lies in its ability to surmount the constraints bedeviling conventional 3D CNNs. These limitations encompass a lack of flexibility and a constrained feature extraction capacity. When juxtaposed with the conventional PND algorithms that dominate the field, our proposed algorithm unfurls its prowess by securing an impressively high CPM score of 0.885, thus establishing its resounding superiority. What's more, our exploration delves deeper as ablation experiments are harnessed to substantiate the manifold contributions of OD3D towards bolstering performance, all while offering an optimal configuration for seamless integration.

Keywords: Pulmonary nodule detection · Dynamic convolution · CT images

1 Introduction

In recent times, an escalating number of researchers have come to acknowledge that lung CT inherently embodies three-dimensional data, encompassing an array of voluminous stereoscopic features across the slice dimension. This realization has led to the development of PND algorithms rooted in three-dimensional convolutional networks (3D CNN), resulting in their demonstration of superior performance. Nonetheless, the restricted adaptability and dynamic feature extraction capability stemming from the weight-sharing nature of CNN networks have emerged as pivotal constraints limiting CNN architecture performance. To

H. Chen et al. (Eds.): TAI4H 2024, LNCS 14812, pp. 13–24, 2024.
https://doi.org/10.1007/978-3-031-67751-9_2

address this issue, researchers have put forward diverse solutions from varying perspectives. Taking inspiration from biological research, certain scholars have devised attention mechanisms, incorporating specific algorithms to empower the network with the ability to focus adaptively on critical features.

This attention mechanism has given rise to a novel category of CNNs, known as dynamic CNNs, which fundamentally reshape traditional static convolutional kernels into dynamic linear combinations of several kernels. In contrast to attention mechanisms that operate on the convolution output layer, dynamic convolution directly influences the convolution kernel weights themselves. This allows the kernels to dynamically adapt to varying data patterns.

Currently, the 3D CNN remains the most mature architecture for PND algorithms. Its simplicity and computational efficiency have cemented its status as a well-established and cost-effective framework for three-dimensional feature extraction. The distinctive advantages offered by dynamic convolution further enrich the insights presented in this work. To more effectively extract features from lung nodules and to advance high-performance PND algorithms, this paper introduces the development of a detection algorithm named Omni-dimension Dynamic Residual 3D Net (ODR3DNet). This network employs an encoder-decoder architecture [6] and leverages full-dimensional dynamic three-dimensional convolution (OD3D) for achieving dynamic feature extraction. The comprehensive three-dimensional network architecture ensures the robust extraction of stereoscopic features.

2   Related Work

The emergence of deep learning (DL) algorithms, particularly the advent of Convolutional Neural Network (CNN) architectures for image-based tasks, has exhibited significant performance advantages, kindling hope for PND algorithms. A scrutiny of current DL-based PND algorithms reveals their classification into two categories: 2D CNN-based two-dimensional detection algorithms and 3D CNN-based three-dimensional detection algorithms.

2D CNN-based algorithms surfaced slightly ahead. Ding et al. [2] introduced a lung nodule detection method based on the Faster R-CNN algorithm, integrating a deconvolutional structure. Setio et al. [14] put forth a multi-view CNN algorithm that amalgamated CT slices from various orientations to generate the final outcome. George et al. [3] presented a 2D CNN regression-based detection algorithm employing the YOLO structure. Gu et al. [4] proposed an improved lung nodule detection model based on deformable convolution, where offsets are obtained by adding a branching network to make the feature extraction process more suitable for the irregular shape of nodules. Nguyen et al. [11] overcame the problem of mismatch between manually designed anchor boxes and the size and shape of lung nodules by designing adaptive anchor boxes.

The aforesaid works brought about substantial enhancements in performance when contrasted with traditional algorithms. Nonetheless, CT images inherently comprise a sequence of slices housing volumetric information. Numerous consecutive slices contain an abundance of 3D information, which is overlooked by 2D

CNNs, rendering them suboptimal. Since the incorporation of 3D CNNs, a growing number of researchers have embraced their considerable advantages. Currently, Li et al. [7] formulated a 3D CNN detection framework named Deepseed, employing an encoder-decoder structure and introducing attention mechanisms to spotlight pivotal lung nodule features. Mei et al. [10] devised the SGNL module, significantly augmenting the long-range dependency acquisition capability of CT images and attaining commendable performance. Zhang et al. [18] obtained an optimized 3D feature pyramid network (FPN) [8] architecture for lung nodule detection by implementing an improvement, which improves the detection performance by adding an SE module with a channel attention mechanism. Zhang et al. [17] designed a deep 3D region suggestion network for detecting lung nodules and a multi-scale feature fusion network for nodule classification.

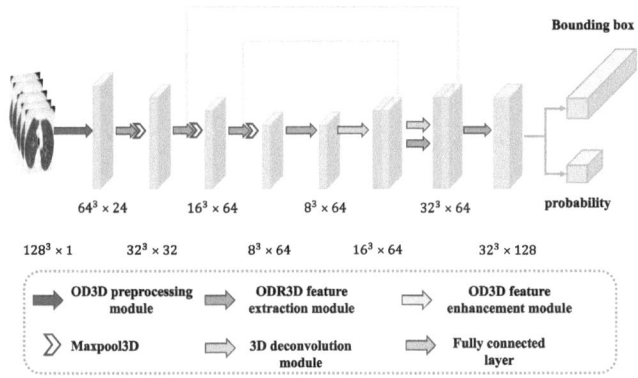

Fig. 1. Overall framework of ODR3DNet. Distinctly colored arrows denote various modules, while the numbers indicate the dimensions of the feature maps (width × height × slice number × channel number). Given that all dimensions are cubes, they can be simplified as the cube edge length cubed × channel number.

3   Method

The overall architecture of the proposed ODR3DNet is visually depicted in Fig. 1. The network adheres to an encoder-decoder framework, where the feature map size progressively reduces in the left encoder segment, facilitating dynamic extraction of lung nodule features via the innovative OD3D dynamic convolution. Subsequent to this, the feature map size is reinstated through a multilevel decoder, ensuring exhaustive recovery of intricate features. Furthermore, by amalgamating feature maps from both the encoder and decoder through skip connections, a more robust feature fusion is accomplished, thereby elevating network performance. The entire algorithm is rooted in 3D CNN design, allowing for the comprehensive capture of stereoscopic features from CT images.

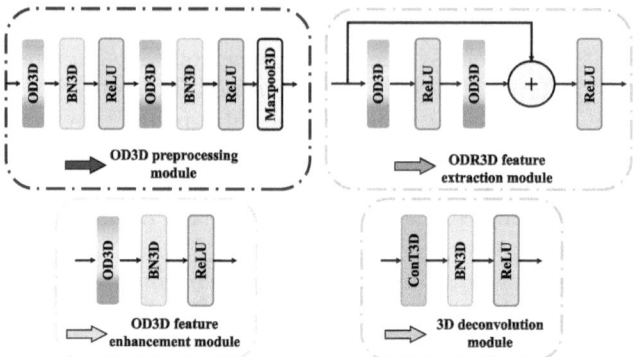

Fig. 2. Detailed structure of ODR3DNet.BN3D corresponds to three-dimensional batch normalization, which is complemented by the ReLU activation function. Maxpool3D signifies 3D max pooling. ConT3D represents the three-dimensional deconvolution module, pivotal for upsampling feature maps within the decoder section.

3.1 Encoder-Decoder Structure

This algorithm extends the concept of residual structures into the three dimensional realm [12], crafting the encoder architecture through the utilization of three-dimensional residual modules. Additionally, this algorithm introduces the full-dimensional dynamic three-dimensional convolution (OD3D) and integrates it within the residual structure, thus empowering the encoder with potent adaptive feature extraction capabilities.

Given that the encoder section extracts abstract features through max pooling, it leads to information loss and generates small-sized feature maps that may not be conducive to subsequent candidate box regression. Consequently, taking a cue from the fundamental principles of U-Net [13], this algorithm designs the decoder framework. The decoder incorporates deconvolutional modules to upscale the feature maps, consequently augmenting the size and resolution of the compact-sized feature maps housing high-level abstract features.

Of utmost significance, the yellow feature-enhancement arrows depicted in Fig. 1 denote the skip connections, facilitating the amalgamation of lower-level, larger-sized feature maps with the decoder's images. This amalgamation enriches the features and significantly amplifies the overall network's detection performance.

3.2 OD3D-Based Feature Extraction Module

In Fig. 1, all the modules featuring OD3D are depicted, encompassing the preprocessing module indicated by the purple arrow, the dynamic residual modules [5] represented by the blue arrow, and the feature enhancement module denoted by the yellow arrow.

The detailed structures of each module are elucidated in Fig. 2. Within the figure, with the exception of the deconvolution module highlighted by the green arrow, the remaining three modules constitute the central components of feature extraction within the encoder.

Independently, Yang et al. [16] and Chen et al. [1] introduced CondConv and DyConv dynamic convolution algorithms respectively. These two pioneering studies hold significance in the domain of dynamic convolution. The core principles underlying both algorithms are largely akin, with structural similarities to Fig. 3 and mathematical expressions akin to Eq. 1:

$$Y = (\alpha_1 W_1 + \alpha_1 W_1 + ... \alpha_N W_N) * X$$
$$= \sum_{i=1}^{N} \alpha_i W_i * X \tag{1}$$

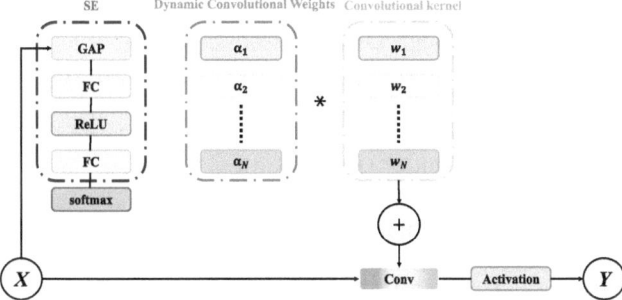

Fig. 3. DyConv. Different convolutional kernels are directly represented by weights.

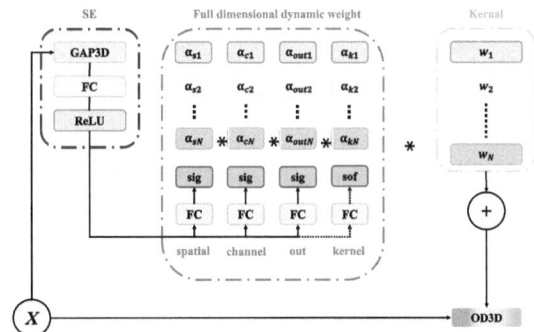

Fig. 4. OD3D.

The single-dimensional dynamic convolution, as portrayed in Eq. 1, comprises two elements: the weight matrix W, representing the convolutional kernel, and the attention matrix α, employed to harmonize the convolutional kernel itself. However, the convolution algorithm's design space is multidimensional, encompassing four dimensions: the kernel's spatial dimension, input channel dimension, output channel dimension, and overall kernel dimension. Moreover, The performance advancement of DyConv is contingent on multiple convolutional kernels, generally with $N = 4$. This implies that widespread DyConv implementation incurs notable computational expenses. In the scenario of a singular kernel ($N = 1$), the attention mechanism itself becomes one-dimensional, largely stripping it of dynamism. Consequently, this dynamic design is undeniably coarse, presenting considerable potential for refinement. So the crux of enhancing dynamic convolution lies in devising more sophisticated dynamic strategies that wholly capture its adaptability and superiority. This is precisely the core concept behind OD3D. OD3D constitutes a completely three-dimensional and all-encompassing dynamic convolution, as depicted in Fig. 4.

Upon comparing Fig. 4 with Fig. 3, it becomes apparent that their fundamental principles align, although OD3D introduces a higher level of complexity. Both methods integrate attention mechanisms derived from the SE mechanism. Nonetheless, OD3D's attention matrix encompasses all four dimensions: the spatial dimension α_s, input channel dimension α_c, output channel dimension α_{out}, and kernel dimension α_k. While the activation functions applied to the first three dimensions are Sigmoid (sig), the kernel dimension's activation function adopts Softmax (sof). Due to OD3D's all-encompassing nature, substantial performance enhancements can be achieved, even with a solitary convolutional kernel ($N = 1$). This stands in marked contrast to DyConv, a distinction that will be further elaborated upon in the subsequent ablation experiment section. In the case of a single convolutional kernel within OD3D, the kernel dimension is disregarded, as indicated by the dashed box in Fig. 4. The formulation of OD3D is articulated in Eq. 2.

$$Y = \left[\sum_{i=1}^{N} (\alpha_{si} \odot \alpha_{ci} \odot \alpha_{\text{outi}} \odot \alpha_{ki}) \odot W_i \right] * X \qquad (2)$$

In the Eq. 2, the spatial dimension attention is denoted as $\alpha_{si} \in \mathbb{R}^{k \times k \times k}$, the input channel dimension attention as $\alpha_{ci} \in \mathbb{R}^{1 \times 1 \times 1 \times C^{in}}$, the output channel dimension attention as $\alpha_{outi} \in \mathbb{R}^{1 \times 1 \times 1 \times C^{out}}$, and the kernel dimension attention as $\alpha_{ki} \in \mathbb{R}^{1 \times 1 \times 1 \times C^n}$.

3.3   Classification Module

In the context of the PND algorithm, once the input data undergoes feature extraction through the encoder-decoder network, the generation of lung nodule detection boxes and their associated classification probabilities is managed by this module, typically situated in the final part of the network. This module is comprised of two parallel and independent fully connected (FC) layers: one dedicated to bounding box regression and the other to classification probabilities.

The bounding boxes adhere to the anchor theory, with this algorithm defining three anchor boxes sized [5, 10, 20] to accommodate varying nodule dimensions. The bounding box FC layer performs regression on the three-dimensional coordinates and diameter of the lung nodules. Simultaneously, the classification FC layer provides the output probability indicating whether the detected object is a lung nodule.

3.4   Loss Functions

The loss function of this algorithm comprises two components: the classification loss L_c and the bounding box regression loss L_r. To tackle the challenge of imbalanced classification samples and to prevent the network from overly focusing on erroneous samples, this algorithm integrates the focal loss [9] into L_c. The classification FC layer of this algorithm generates the probability of the detected object being a lung nodule, represented as P. The formulation for L_c is provided below:

$$L_c = -\alpha \left(1 - P_t\right)^{\gamma} \log\left(P_t\right) \tag{3}$$

In the Eq. 3, if the target region is indeed a lung nodule, then $P_t = P$; otherwise, it indicates a classification error and $P_t = 1 - P$. α and γ are harmonic hyperparameters, and in this algorithm, they are set as $\alpha = 0.5$ and $\gamma = 2$.

The bounding box regression loss, L_r, is defined as follows:

$$L_r = \sum_k L1(G_k, Pk) \tag{4}$$

In Eq. 4, $L1(.)$ denotes the L1 regularization function, and G_k and P_k represent the ground truth relative coordinates and predicted relative coordinates, respectively. They are defined as follows:

$$
\begin{aligned}
G_k &= \left(\frac{x_g - x_\alpha}{r_\alpha}, \frac{y_g - y_\alpha}{r_\alpha}, \frac{z_g - z_\alpha}{r_\alpha}, \log\frac{r_g}{r_\alpha}\right) \\
P_k &= \left(\frac{x - x_\alpha}{r_\alpha}, \frac{y - y_\alpha}{r_\alpha}, \frac{z - z_\alpha}{r_\alpha}, \log\frac{r}{r_\alpha}\right)
\end{aligned}
\tag{5}
$$

(x_g, y_g, z_g, r_g) signifies the actual coordinates of the label, whereas (x, y, z, r) stands for the coordinates predicted by this algorithm. $(x_\alpha, y_\alpha, z_\alpha, r_\alpha)$ represents the anchor coordinates, and k denotes the batch size during training. Hence, by amalgamating Eq. 4 through Eq. 5, the model continually adjusts the predicted coordinates based on the actual values.

The conclusive loss function of this algorithm, labeled as L, amalgamates L_c and L_r, defined as $L = L_c + \delta L_r$. Only when the predicted outcome is a positive sample (lung nodule), $\delta = 1$, indicating the necessity for further bounding box regression. In contrast, when $\delta = 0$, the bounding box regression operation is bypassed directly.

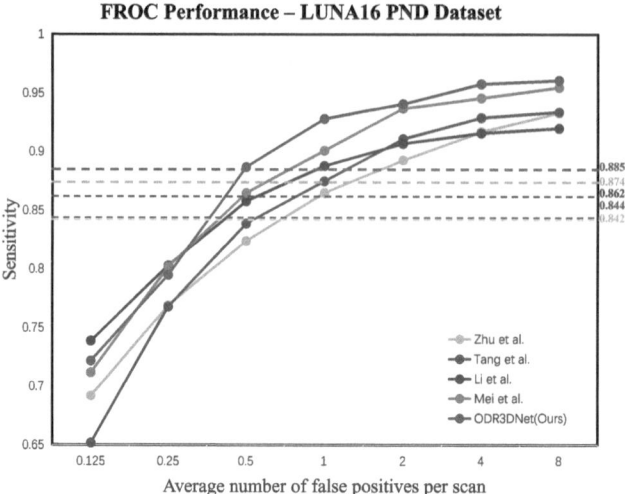

Fig. 5. FROC diagram of mainstream PND algorithm.

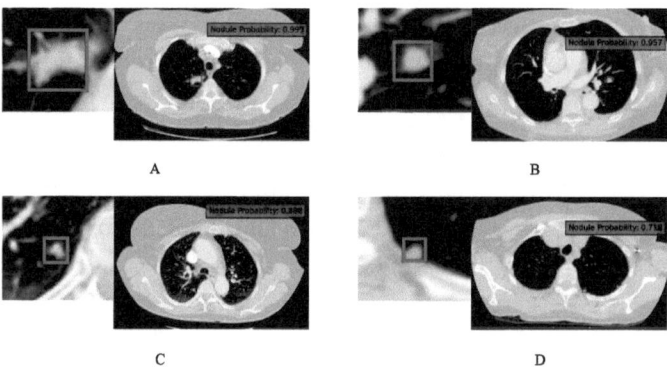

Fig. 6. Visualization of ODR3DNet detection results.

Table 1. Performance comparison of mainstream PND algorithms

FPs/scan	0.125	0.25	0.5	1	2	4	8	CPM
Zhu et al.	0.692	0.769	0.824	0.865	0.893	0.917	0.933	0.842
Tang et al.	0.652	0.768	0.839	0.875	0.911	0.929	0.934	0.844
Li et al.	**0.739**	**0.803**	0.858	0.888	0.907	0.916	0.920	0.862
Mei et al.	0.712	0.802	0.865	0.901	0.937	0.946	0.955	0.874
ODR3DNet	0.722	0.795	**0.887**	**0.928**	**0.941**	**0.958**	**0.961**	**0.885**

Table 2. Number of convolutional kernal in OD3D module(*is best)

FPs/scan	0.125	0.25	0.5	1	2	4	8	CPM	GPU Memory
3D CNN	0.682	0.775	0.821	0.865	0.886	0.908	0.934	0.839	11.768
1	0.709	0.787	0.868	0.896	0.921	0.932	0.947	0.866	13.463
2*	0.722	0.795	0.887	0.928	0.941	0.958	0.961	0.885	22.587
4	**0.735**	**0.801**	**0.897**	**0.934**	**0.946**	**0.965**	**0.971**	**0.893**	41.268

Table 3. Contribution of OD3D full-dimensional dynamic three-dimensional convolution

FPs/scan	0.125	0.25	0.5	1	2	4	8	CPM
3D CNN	0.682	0.775	0.821	0.865	0.886	0.908	0.934	0.839
DyConv [6]	0.714	0.781	0.835	0.872	0.904	0.918	0.936	0.851
OD3D*	**0.722**	**0.795**	**0.887**	**0.928**	**0.941**	**0.958**	**0.961**	**0.885**

4 Experiments

4.1 Dataset

The experimental Pulmonary Nodule Detection (PND) dataset utilized in this study is LUNA16 [15]. Due to the computational intricacies of the 3D CNN, it is neither practical nor meaningful to directly feed the fully processed CT files into the network for training. Therefore, our approach selectively extracts local CT data with dimensions of $128 \times 128 \times 128$ (height $\times$ width $\times$ number of slices) from each patient for network training. During network testing, larger image sizes are extracted to emulate the clinical scenario ($208 \times 208 \times 208$). Given the dataset's limited size, essential data augmentation techniques are employed to enhance training performance. These strategies encompass random cropping, random flipping, and random scaling (within a scaling range of $[0.75, 1.25]$).

4.2 Experimental Setting

The algorithm utilizes the stochastic gradient descent (SGD) optimizer for training, implementing a dynamic learning rate strategy. The initial learning rate is established at 0.01, and the training spans 150 epochs. After the completion of 100 epochs, the learning rate is diminished to 0.001, with a further reduction to 0.0001 occurring after 120 epochs. The batch size is set to 8, and the codebase is developed using Python 3.7 alongside PyTorch 1.6.0. Algorithmic training and experimentation are executed on a high-capacity server, featuring four NVIDIA TITAN GPUs (12GB each). The underlying operating system is Ubuntu 20.04, with GPU acceleration facilitated through CUDA 10.0.

To conform to the stipulations of LUNA16, the algorithm adopts a tenfold cross-validation approach during both training and validation phases. This ensures the attainment of stable experimental outcomes and bolsters model resilience. Evaluation metrics encompass the free-response receiver operating characteristic (FROC) curve and the competitive performance measure (CPM). In this context, sensitivity stands as a crucial parameter, quantified using the subsequent equation:

$$Sensitivity = \frac{TP}{TP + FN} \qquad (6)$$

In Eq. 6, TP signifies the count of accurately predicted lung nodules, whereas FN represents the tally of true lung nodules mistakenly categorized as false positives.

Adhering to the endorsed evaluation criteria, derivation of the free-response receiver operating characteristic (FROC) curve involves computing sensitivity at distinct false positive rates (FPs/scan), encompassing a spectrum from 0.125 to 8 per CT image. The competitive performance measure (CPM) stands as an averaged sensitivity across seven pre-defined average false positive rates: 1/8, 1/4, 1/2, 1, 2, 4, and 8.

4.3   Comparative Experiment and Analysis

In this section, we provide a detailed comparative analysis of the performance of current state-of-the-art PND algorithms. All algorithms were trained on the LUNA16 dataset, and the experimental results are presented in Table 1. The corresponding FROC curves are shown in Fig. 5. The experimental results clearly demonstrate the superiority of the proposed ODR3DNet, a full-dimensional residual dynamic convolutional network, which achieves the highest CPM performance among the compared algorithms.

To visually demonstrate the performance of our algorithm, Fig. 6 shows the visualization results of four cases. As shown in the figure, for each case, the red bounding box on the left is an enlargement of the red bounding box on the right side of the main image, and the predicted probability is shown in the upper right corner of the main image. The predictive probabilities and bounding boxes reflect the accuracy of our algorithms. Our algorithm is able to accurately detect lung nodules of different shapes and sizes. Even in challenging cases like the one depicted in Fig. 6(D), where the nodules are small and difficult to detect, our algorithm can still achieve accurate detection. While the prediction probability may be relatively low, it still demonstrates the robust detection capabilities of ODR3DNet.

4.4   Ablation Experiment and Analysis

For ODR3DNet, the utmost critical hyperparameter is the count of convolutional kernels, denoted as "n" within each dynamic convolution module. The core constituent is the OD3D full-dimensional dynamic convolution. The outcomes of the experiments are detailed in Tables 2 and 3.

Table 2 demonstrates the repercussions of varying the numbers of convolutional kernels in the OD3D dynamic convolution module. With the intent of lucidly showcasing the ascendancy of dynamic convolution, we intentionally substituted all OD3D occurrences with standard 3D CNN for juxtaposition. The findings divulge that the utmost performance occurs when n = 4, albeit at a significant memory expense, delivering only marginal enhancements relative to n = 2. The juxtaposition against 3D CNN attests that even with a sole convolutional kernel, the full-dimensional dynamic mechanism still results in notable performance strides. Opting for n = 2 accomplishes veritable "full-dimensionality" while upholding a well-balanced performance, positioning it as the optimal choice.

Table 3 illustrates the limitations of regular static convolution compared to dynamic convolution. By replacing the 3D CNN structure with DyConv, the performance improvement brought by dynamic convolution is reflected. However, OD3D demonstrates greater flexibility, adaptability, and dynamic nature compared to the single-dimensional DyConv structure, resulting in better performance.

5   Conclusion

The ODR3DNet adaptive PND algorithm is entirely crafted upon a three dimensional convolutional neural network, profoundly capable of extracting the stereoscopic attributes of lung nodules. Within this algorithm, a comprehensive full-dimensional dynamic three-dimensional convolution module, OD3D, is meticulously fashioned to supplant conventional three-dimensional convolutions. OD3D adeptly harnesses all dimensions of the convolutional kernel, orchestrating dynamic feature extraction across the entire spectrum, thereby furnishing the algorithm with potent adaptive feature extraction capabilities. In comparison to prevalent PND algorithms, this algorithm attains an exceptional CPM score of 0.885, a testament to its substantial superiority. Ablation experiments corroborate OD3D's instrumental role in performance enhancement and validate the final configuration's merits.

References

1. Chen, Y., Dai, X., Liu, M., Chen, D., Yuan, L., Liu, Z.: Dynamic convolution: attention over convolution kernels. In: Proceedings of the IEEE/CVF Conference on Computer Vision and Pattern Recognition, pp. 11030–11039 (2020)
2. Ding, J., Li, A., Hu, Z., Wang, L.: Accurate pulmonary nodule detection in computed tomography images using deep convolutional neural networks. In: Descoteaux, M., Maier-Hein, L., Franz, A., Jannin, P., Collins, D.L., Duchesne, S. (eds.) MICCAI 2017. LNCS, vol. 10435, pp. 559–567. Springer, Cham (2017). https://doi.org/10.1007/978-3-319-66179-7_64
3. George, J., Skaria, S., Varun, V., et al.: Using yolo based deep learning network for real time detection and localization of lung nodules from low dose CT scans. In: Medical Imaging 2018: Computer-Aided Diagnosis, vol. 10575, pp. 347–355. SPIE (2018)

4. Gu, J., Tian, Z., Qi, Y.: Pulmonary nodules detection based on deformable convolution. IEEE Access **8**, 16302–16309 (2020). https://api.semanticscholar.org/CorpusID:210972598

5. He, K., Zhang, X., Ren, S., Sun, J.: Deep residual learning for image recognition. In: Proceedings of the IEEE Conference on Computer Vision and Pattern Recognition, pp. 770–778 (2016)

6. Li, C., Zhou, A., Yao, A.: Omni-dimensional dynamic convolution. arXiv preprint arXiv:2209.07947 (2022)

7. Li, Y., Fan, Y.: Deepseed: 3D squeeze-and-excitation encoder-decoder convolutional neural networks for pulmonary nodule detection. In: 2020 IEEE 17th International Symposium on Biomedical Imaging (ISBI), pp. 1866–1869. IEEE (2020)

8. Lin, T.Y., Dollár, P., Girshick, R.B., He, K., Hariharan, B., Belongie, S.J.: Feature pyramid networks for object detection. In: 2017 IEEE Conference on Computer Vision and Pattern Recognition (CVPR), pp. 936–944 (2016). https://api.semanticscholar.org/CorpusID:10716717

9. Lin, T.Y., Goyal, P., Girshick, R., He, K., Dollár, P.: Focal loss for dense object detection. In: Proceedings of the IEEE International Conference on Computer Vision, pp. 2980–2988 (2017)

10. Mei, J., Cheng, M.M., Xu, G., Wan, L.R., Zhang, H.: Sanet: a slice-aware network for pulmonary nodule detection. IEEE Trans. Pattern Anal. Mach. Intell. **44**(8), 4374–4387 (2021)

11. Nguyen, C.C., Tran, G.S., Nguyen, V.T., Burie, J.C., Nghiem, T.P.: Pulmonary nodule detection based on faster R-CNN with adaptive anchor box. IEEE Access **9**, 154740–154751 (2021). https://api.semanticscholar.org/CorpusID:244419436

12. Ren, S., He, K., Girshick, R., Sun, J.: Faster R-CNN: towards real-time object detection with region proposal networks. In: Advances in Neural Information Processing Systems, vol. 28 (2015)

13. Ronneberger, O., Fischer, P., Brox, T.: U-Net: convolutional networks for biomedical image segmentation. In: Navab, N., Hornegger, J., Wells, W.M., Frangi, A.F. (eds.) MICCAI 2015. LNCS, vol. 9351, pp. 234–241. Springer, Cham (2015). https://doi.org/10.1007/978-3-319-24574-4_28

14. Setio, A.A.A., et al.: Pulmonary nodule detection in CT images: false positive reduction using multi-view convolutional networks. IEEE Trans. Med. Imaging **35**(5), 1160–1169 (2016)

15. Setio, A.A.A., et al.: Validation, comparison, and combination of algorithms for automatic detection of pulmonary nodules in computed tomography images: the luna16 challenge. Med. Image Anal. **42**, 1–13 (2017)

16. Yang, B., Bender, G., Le, Q.V., Ngiam, J.: Condconv: conditionally parameterized convolutions for efficient inference. In: Advances in Neural Information Processing Systems, vol. 32 (2019)

17. Zhang, H., Zhang, H.: Lungseek: 3D selective kernel residual network for pulmonary nodule diagnosis. Vis. Comput. **39**, 679 – 692 (2022). https://api.semanticscholar.org/CorpusID:246359391

18. Zhang, M., Kong, Z., Zhu, W., Yan, F., Xie, C.: Pulmonary nodule detection based on 3d feature pyramid network with incorporated squeeze-and-excitation-attention mechanism. Concurr. Comput. Pract. Exp. **35** (2021). https://api.semanticscholar.org/CorpusID:233638919

Knowledge Injected Multimodal Irregular EHRs Model for Medical Prediction

Sicen Liu and Hao Chen$^{(\boxtimes)}$

The Hong Kong University of Science and Technology, Kowloon, Hong Kong
`jhc@cse.ust.hk`

Abstract. The health conditions among patients in intensive care units (ICUs) are essential for providing optimal care and improving patient outcomes. Electronic health records (EHRs) serve as a valuable source of information in ICUs, capturing numerical time series data and lengthy clinical note sequences. However, these EHRs often present challenges due to the irregularity of data collection at varying time intervals. Additionally, existing works ignored to incorporate domain knowledge and integrate memory knowledge from existing multimodal EHRs to improve the care process. Our method first addresses the challenges by (1) transforming clinical note representations into multivariate irregular time series, where each note is associated with different time intervals; (2) injecting static domain knowledge based on patient clinical manifestations into patient representations; (3) storing knowledge from the samples in dynamic memory and learning to inject memory knowledge based on patient representations. The exceptional performance of our proposed methods in two medical prediction tasks consistently outperforms state-of-the-art (SOTA) baselines in both single modality and multimodal fusion scenarios.

Keywords: Medical Predictions · Knowledge Injection · Irregular Multimodal Electronic Health Records

1 Introduction

Intensive Care Units (ICUs) serve as critical care settings for patients afflicted with life-threatening conditions, such as trauma [1], sepsis [2], and organ failure [3]. The initial hours following admission play a pivotal role in determining patient outcomes, while also presenting a higher susceptibility to medical decision errors compared to later stages [4,5]. To assist clinicians in delivering appropriate treatments, the integration of automated tools capable of providing effective and real-time predictions holds significant promise. Notably, the advent of Electronic Health Records (EHRs) has facilitated the recording of patients' health conditions within ICUs [6–8], thereby opening opportunities for the application of deep neural networks in healthcare [9–11]. These applications encompass diverse areas, including mortality prediction [12,13], phenotype classification [14], medication

H. Chen et al. (Eds.): TAI4H 2024, LNCS 14812, pp. 25–39, 2024.
https://doi.org/10.1007/978-3-031-67751-9_3

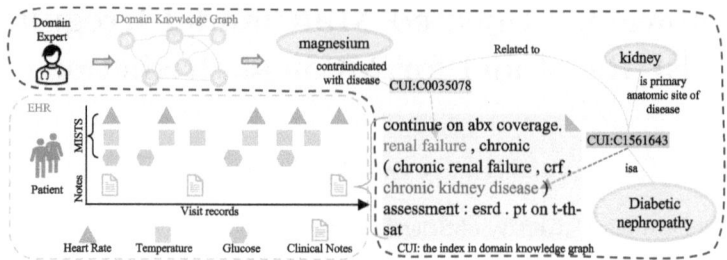

Fig. 1. An example of patient admission to the ICU involves MISTS (i.e., heart rate, temperature, and glucose levels) and a series of clinical notes in the EHR. The clinical records are typically unstructured text, and the collection of time points is considerably sparser compared to clinical measurements. On the other hand, the domain knowledge graph comprises clinical facts that have been synthesized by domain experts. Here, we showcase the external knowledge of various entities present in the patient's clinical notes as represented within the domain knowledge graph.

recommendation [15, 16]. EHRs encompass multivariate irregularly sampled time series (MISTS) and irregular clinical note sequences. Leveraging domain-specific knowledge graphs curated by domain experts, we can provide specialized domain knowledge for coarse-grained irregular clinical record sequences, as shown in Fig. 1. In the provided Fig. 1, it is evident that there are several crucial patient manifestation entities within the clinical notes. Merely relying on the information within the clinical notes alone falls short of capturing a more comprehensive understanding of domain-specific factual knowledge. For instance, when considering the entity *chronic kidney disease* and incorporating domain knowledge, we not only acquire the knowledge that it *is the primary anatomic site of disease* in the *kidney* but also understand its association with *renal failure*. These insights cannot be obtained solely from the patient's clinical notes. Therefore, effectively integrating domain knowledge allows us to obtain personalized patient representations with the guidance of specialized information. However, previous works in medical prediction have often overlooked the importance of domain-specific factual knowledge. Additionally, patients' samples also contain unique knowledge information, but prior works have typically treated patients as separate individuals during training, neglecting the interconnected knowledge among patients. The multimodal structure, intricate irregular temporal properties, and the prior neglect of incorporating domain knowledge have posed challenges to prediction tasks. Consequently, we formulate three research objectives:

- Tackling irregularity in both clinical note sequence and MISTS.
- Incorporating static domain knowledge information into the learning process of patients' irregular multimodal representations.
- Leveraging dynamic memory mechanisms to integrate the knowledge stored in samples and inject it into the patient's representation.

To address the aforementioned research objectives, we propose the Knowledge Injected Multimodal Irregular EHR model for medical prediction, named as

KIMP. Specifically, for research objective one, we introduce a multimodal representation learning approach using irregular temporal encoding for both MISTS and clinical notes. For research objective two, we first extract key entities from the lengthy text sequences in clinical notes and map them to domain knowledge, thereby obtaining a subgraph of the domain knowledge graph. This process leads to a clinical information representation enhanced by domain knowledge. For research objective three, we design a dynamic memory network to automatically learn the knowledge information stored within the samples themselves. Compared to the baseline methods, our approach demonstrates excellent performance in both unimodal and multimodal fusion scenarios.

2   Related Work

Irregular Electronic Health Records Modeling. Irregular Electronic Health Records (EHRs) encompass two primary components: Multimodal Irregular Time Series (MISTS) and clinical notes. MISTS, which stands for observations acquired at irregular time intervals with potential misalignments across variables [17]. GRU-D [18] utilizes gated recurrent units to capture temporal dependencies by decaying hidden states. RAINDROP [19] treats MISTS as separate sensor graphs, leveraging graph neural networks to learn variable dependencies. These methods offer specialized designs that capture irregular temporal dependencies in MISTS from distinct perspectives. However, Temporal Discretization and Embedding (TDE) methods, which convert MISTS into fixed-dimensional feature spaces and utilize deep neural models for regular time series, represent a subset of approaches for handling MISTS. To address this gap, [20] introduce interpolation-prediction networks (IP-Nets) that utilize a kernel function with learned parameters to interpolate MISTS at regular reference points. Building upon this work, [21] presents a time attention mechanism with time embeddings to learn interpolation representations, further enhancing the ability to handle irregular temporal dependencies in MISTS. UTDE [22] leverages different TDE methods as submodules and integrates hand-crafted imputation embeddings into learned interpolation embeddings to improve medical predictions. For irregular clinical note modeling, [23] and [24] concatenate the patient's clinical notes, and subsequent utilization of BERT variants [25,26] to obtain text representations. Notably, these approaches overlook the inherent irregularity present within clinical notes. To address this limitation, [27] propose a time-awarded LSTM with a trainable decay function, which aims to capture the irregular time information associated with clinical notes. However, the effectiveness of this approach may be constrained by the presence of limited parameters. To comprehensively account for the irregularity inherent in clinical notes, UTDE [22] cast the representations of clinical notes, considering their irregular note-taking times, as MISTS. This formulation allows each dimension of a series of clinical note representations to be treated as an irregular time series.

Knowledge Enhanced Electronic Health Records Modeling. In healthcare, incorporating domain knowledge has become a common practice to gain a better understanding of patients' health conditions and improve medical predictions. Integrating domain knowledge into medical prediction tasks enables models to consider relevant factors and features that are specific to the healthcare domain, ultimately leading to more accurate and reliable predictions. For example, KAME [28] and GRAM [29], have leveraged medical knowledge graphs with attention mechanisms on ICD codes to facilitate multi-disease diagnosis. By assigning attention weights to the nodes of the knowledge graph, these models can capture the diseases experienced by a patient during each hospital visit. CAML [30] and LDAM [31] incorporate external medical information through the cross-attention mechanism in models. These approaches utilize target disease risk labels to guide the attention of the disease diagnosis model toward medical features relevant to the specific target label. In this paper, we incorporate fine-grained domain knowledge derived from the Unified Medical Language System (UMLS) [32] for precision medical prediction. The UMLS is a comprehensive and widely used domain knowledge resource that contains various biomedical vocabularies, ontologies, and terminologies and relationship between them. These fine-grained domain knowledge representations provide profession insights into the intricate relationships within the clinical notes. Additionally, prior research has commonly treated patients as isolated individuals during the training process, consequently neglecting the interconnected knowledge that can be gleaned from similar patients. Our research highlights the limitations of previous approaches and emphasizes the need to incorporate domain-specific factual knowledge and consider shared knowledge among patients.

3   Method

Our proposed model comprises three primary components: Domain Knowledge Injection, In-Memory Knowledge Injection, and Cross-Modal Fusion, as shown in Fig. 2. In this section, we will illustrate each part thoroughly.

3.1   Problem Setup

Denote $\mathcal{D} = \{(x_i^m, t_i^m), (x_i^n, t_i^n), y_i\}_{i=1}^{\mathbb{N}}$ to be an EHR dataset with $\mathbb{N}$ patients, where (x_i^m, t_i^m) is d_m-dimensional MISTS, x_i^m being observations and t_i^m being corresponding time points, (x_i^n, t_i^n) is a series of clinical notes with note-taking time and y_i is the target outcome, e.g. discharge or death for modality prediction. For simplicity, we omit the patient index i in the subsequent section. Each dimension of the MISTS, (x_j^m, t_j^m), where $j = \{1, \cdots, d_m\}$, has l_j^m observations, and each patient's (x^n, t^n) includes l^n clinical notes. We represent the domain knowledge as a graph denoted by $\mathcal{G} = \{(h, r, t)\}$, where h is the head entity and t is the tail entity, the r is the relationship between head entity and tail entity. In early-stage medical predictions, given (x_i^m, t_i^m) and (x_i^n, t_i^n) before a certain time point (e.g. 48-hour) after admission, and domain knowledge $\mathcal{G}$, we seek to predict y for every patient.

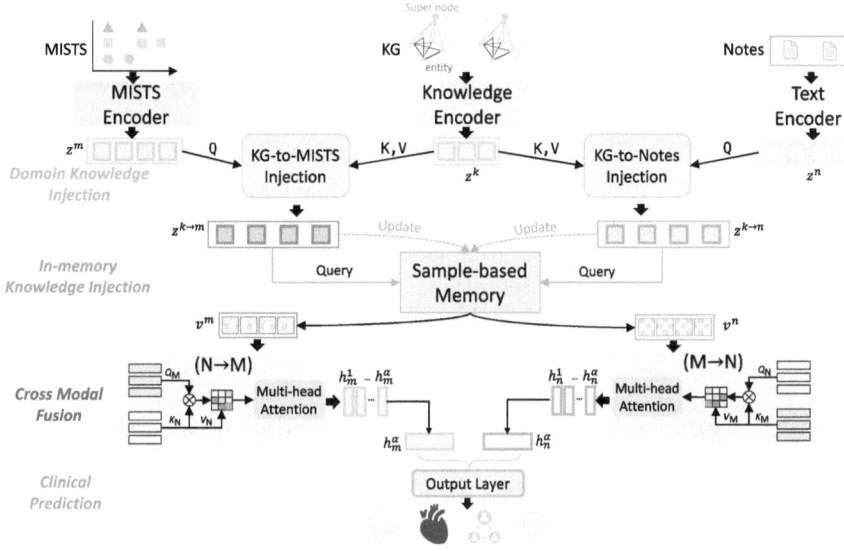

Fig. 2. The overview of our proposed KIMP involves taking MISTS, clinical notes, and domain knowledge as inputs. The multimodal features then pass through the domain knowledge injection, in-memory knowledge injection, and cross-modal fusion modules. Finally, the classifier predicts clinical outcomes based on these processed features.

3.2   Domain Knowledge Injection

MISTS Encoder. To incorporate irregular time information of MISTS, we follow UTDE [22] to obtain MISTS embedding z^m. Specifically, we first discretize x^m based on t^m, to hourly time intervals with a sequence of regular time points, $\alpha = [0, 1, \cdots, \alpha - 1]$. For example, given the time point $\alpha = [0, 1, 2, 3]$ for the first 4 h, the observed features collected at $[1.2, 1.5, 3.7]$ hours post-admission, with values of $[10, 8, 12]$, will be discretized as $[miss1, 8, miss2, 12]$, where $miss1$ and $miss2$ will be imputed using the global mean and the previous observation, respectively. We then fed regular time series into a $1D$ causal convolutional layer with stride 1 to obtain imputation embedding $e^{m_{imp}} \in \mathbb{R}^{\alpha \times d_h}$. Secondly, a time representation, Time2Vec [33], is learned to transform each value in a list of continuous time points and obtain discretized multi-time embeddings $e^{m_{attn}} \in \mathbb{R}^{\alpha \times d_h}$. Finally, a gating function is used to dynamically integrate $e^{n_{imp}}$ into $e^{m_{attn}}$ to obtain compounding embeddings z^m. Formally,

$$z^m = \mathbf{g} \odot e^{m_{imp}} + (1 - \mathbf{g}) \odot e^{m_{attn}} \tag{1}$$

$$e^{m_{attn}} = Attn(\theta_{T2V}(\alpha)w_m^q, \theta_{T2V}(t^m)w_m^k, x^m) \tag{2}$$

$$\theta_{T2V}[\tau] = \begin{cases} w_\tau \tau + \phi_i & \text{if } i = 1 \\ \sin(w_\tau \tau + \phi_i), & \text{if } 1 < i \leq d_v \end{cases} \tag{3}$$

where $\mathbf{g}$ is a gating function implemented by MLP for simplicity and $\odot$ is point wise multiplication. Att is multi-head attention [34] and w_m^q and w_m^k are learned parameters. τ is a list of continuous time points and d_v is vector dimension.

Text Encoder. To retrieve pertinent knowledge from the clinical notes, we begin by utilizing a $TextEncoder$, which is a language model pretrained specifically for the medical domain, to encode the notes. Subsequently, we extract the representation of the [CLS] token from each encoded clinical note. This process yields a sequence of note representations, denoted as $e^n \in \mathbb{R}^{l^n \times d_n}$, where d_n represents the hidden dimension of the encoded text. Formally,

$$e^n = TextEncoder(x^n) \tag{4}$$

In order to address irregularity, we employ a sorting mechanism on the x^n representations based on t^n, resulting in a structure known as MISTS (x^m, t^m). This sorting procedure ensures that each hidden dimension of x^n forms a time series sequence, with all time series sequences sharing the same collection of time points. By organizing the data in this manner, we can effectively handle irregularities and enable consistent analysis and modeling of the time-dependent information contained within the x^n representations.

$$z^n = Attn(\theta_{T2V}(\alpha)w_n^q, \theta_{T2V}(t^n)w_n^k, e^n) \tag{5}$$

where w_n^q and w_n^k are learnable parameters.

Knowledge Encoder. Domain knowledge refers to factual information derived by domain experts within a specific field. To effectively capture the representation of static domain knowledge, we employ a two-step process for knowledge representation. Firstly, we utilize a knowledge representation learning algorithm, such as TransE [35], to encode the knowledge graph $\mathcal{G} = \{k_i = (h_i, r_i, t_i)\}_{i=1}^{N_g}$ to obtain the entity embeddings $\{e_i\}_{i=1}^{N_e}$, where $e_i \in \mathbb{R}^{d_e}$, d_e is the dimension of the entity embeddings. This algorithm enables us to capture the semantic relationships between entities in the knowledge graph. Secondly, we leverage the power of Graph Neural Networks (GNNs), specifically Graph Attention Networks [36], to consider the complete graph structure. By aggregating local information from the graph neighborhood for each node, GNNs allow us to obtain comprehensive entity representations that encapsulate the collective knowledge encoded within the graph. In order to effectively integrate domain knowledge with patient representations, we first extract entities $\{\mathcal{E}_g\}$ from clinical notes x_j^n at time point t_j^n, then the entities $\{\mathcal{E}_g\}$ serve as anchor points for incorporating domain knowledge. This process allows us to construct domain knowledge subgraphs that are relevant to these entities. Additionally, to consolidate the subgraphs obtained from the retrieval of entities, we introduce a *Super node* $\mathcal{E}_s$ that merges these subgraphs into a larger subgraph representing the current clinical notes. This innovative methodology facilitates the seamless integration of domain knowledge with patient representations, enabling a comprehensive understanding of the health condition of the patient.

$$z^k = [e_1^k \oplus e_2^k \oplus \cdots \oplus e_{l^n}^k] \tag{6}$$

$$e_j^k = \mathbf{MH}([e_{js} \oplus \{e_{jg}\}]) \tag{7}$$

where e_{js} is the super node embedding of the j-th knowledge subgraph, $\{e_{jg}\}$ is the collection of embeddings for all entities within the j-th knowledge subgraph. The $\oplus$ is concatenate operation, and $\mathbf{MH}$ is multi-head self attention mechanism [34].

KG Injection. After obtaining the representations of MISTS, clinical notes, and the knowledge graph, a key challenge lies in effectively injecting domain knowledge into the MISTS and clinical notes representations for the patients. To address this, we propose two modules specifically designed for injecting static domain knowledge: the KG-to-MISTS injection submodule and the KG-to-Notes injection submodule. These modules play a crucial role in seamlessly integrating the relevant domain knowledge into the MISTS and clinical notes representations. By leveraging these modules, we can enhance the overall understanding of the patient's health condition by incorporating valuable insights from the domain knowledge graph. Here two mulit-head cross-attention (**CMH**) [37] are leveraged to inject domain knowledge into MISTS and clinical notes. Specifically, for the KG-to-MISTS injection module, we takes the MISTS representation z^m as queries, domain knowledge graph representation z^k as keys and values. For the KG-to-Notes injections module, we takes the notes representation z^n as queries, z^k as keys and values. Therefore, two domain knowledge enhanced MISTS and notes representation is obtained by

$$z^{k \to m} = \mathbf{CMH}_m(z^m, z^k) \tag{8}$$

$$z^{k \to n} = \mathbf{CMH}_n(z^n, z^k) \tag{9}$$

where $\mathbf{CMH}_*, * \in \{m, n\}$ is multi-head cross-attention mechanism [37].

3.3   In-memory Knowledge Injection

In clinical practice, doctors interact with a diverse range of patients, each presenting unique medical conditions and histories. With the progressive accumulation of patient visits, doctors acquire valuable clinical experience that contributes to their expertise. Nonetheless, traditional methodologies often fail to fully exploit the sample-based knowledge that progressively grows alongside the expanding patient sample size. To overcome this limitation, we propose the incorporation of an innovative In-memory knowledge injection module into our framework. This module serves as a repository for storing and retrieving knowledge derived from previous patient samples, dynamically adapting to the evolving clinical landscape. By effectively injecting this memory-based knowledge into patient representations, our model becomes equipped with a comprehensive understanding of various medical scenarios, leading to improved decision-making capabilities and enhanced performance in real-world clinical settings. Specifically, we introduce a sample-based memory slot M_s to dynamically store and retain knowledge obtained during the training process. This memory slot serves as a repository for capturing crucial sample-based knowledge, continually updated based on the input multimodal patient data. By leveraging this mechanism, we are able to extract and preserve valuable insights gained from previous patient encounters, enhancing the model's capability to make informed decisions. Furthermore, we employ the patient's current visit representation to effectively inject the sample-based knowledge into the decision-making process. This injection mechanism ensures that the knowledge acquired from previous patient visits is seamlessly integrated into the current patient's health condition, empowering the our method to learn in-memory knowledge enhanced patient representation.

$$v^m = Attn(z^{k \to m}, M_s, M_s) \tag{10}$$

$$v^n = Attn(z^{k \to n}, M_s, M_s) \tag{11}$$

$$M_s = \mathbf{CMH}_s(M'_s, [z^{k \to m} \oplus z^{k \to n}]) \tag{12}$$

where M'_s is the previous sample-based memory slot, $\mathbf{CMH}_s$ is the multi-head cross-attention mechanism to update the sample-based memory representation, $\oplus$ is the concatenate operation.

3.4   Cross Modal Fusion

After obtaining the representations enhanced with domain knowledge and sample-based memory, it becomes crucial to explore the fusion of information between these two distinct modalities. To address this, we introduce our cross-modal fusion module. Specifically, we leverage two multi-head cross-attentions to capture the cross-modal information between the modalities. These multi-head cross-attentions mechanisms (CMH) enable the learning of knowledge from one modality attended by the other modality, and vice versa. By employing CMH, we obtain outputs that capture the contextual interdependencies between the modalities. Subsequently, we apply a multi-head self-attention mechanism (MH) to each modality's output, allowing us to derive contextual unimodal representations. This approach facilitates the fusion of information and the extraction of essential features from each modality, thereby enabling a comprehensive understanding of the multimodal patient data. Formally,

$$\{h_m^1, \cdots, h_m^\alpha\} = \mathbf{MH}(\mathbf{CMH}(v^m, v^n)) \tag{13}$$

$$\{h_n^1, \cdots, h_n^\alpha\} = \mathbf{MH}(\mathbf{CMH}(v^n, v^m)) \tag{14}$$

where $\{h_*^1, \cdots, h_*^\alpha\}, * \in \{m, n\}$ are the hidden representation of the MISTS and clinical note for the α time point.

3.5   Clinical Prediction

After obtaining the hidden representations of the patients, we follow the previous works [22], which choose to utilize only the hidden state representation from the final time step for clinical prediction.

$$y'_i = f_i(w_i[h_m^\alpha \oplus h_n^\alpha] + b_i) \tag{15}$$

where f_i is a classifier with fully-connected layers to make predictions, w_i and b_i are learnable parameters.

Table 1. Comparison of KIMP and baseline methods. We report average performance on the five random seeds, with standard deviation as the subscripts. The **best** and 2nd best methods under each setup are bold and underlined, respectively. $\mathcal{M}$ represents MISTS, $\mathcal{N}$ represents clinical notes, and $\mathcal{K}$ represents domain knowledge.

Modality	Method	IHM			PHE		
		$F1_{std.}$	$AUPR_{std.}$	$AUROC_{std.}$	$F1_{std.}$	$AUPR_{std.}$	$AUROC_{std.}$
$\mathcal{M}$	mTAND	$45.03_{0.70}$	$47.19_{1.31}$	$84.61_{0.35}$	$27.15_{0.46}$	$39.47_{0.20}$	$74.42_{0.14}$
$\mathcal{M} \times \mathcal{K}$	mTAND$_{\mathcal{M},\mathcal{K}}$	$50.72_{1.56}$	$54.18_{1.48}$	$87.55_{0.29}$	$40.54_{0.21}$	$50.11_{0.19}$	$79.35_{0.15}$
$\mathcal{N}$	mTAND	$52.56_{0.43}$	$55.71_{0.35}$	$88.48_{0.28}$	$51.79_{0.08}$	$58.85_{0.04}$	$83.01_{0.08}$
$\mathcal{N} \times \mathcal{K}$	mTAND$_{\mathcal{N},\mathcal{K}}$	$55.49_{1.11}$	$58.94_{0.76}$	$89.26_{0.14}$	$53.18_{0.25}$	$58.99_{0.09}$	$83.17_{0.06}$
$\mathcal{M} \times \mathcal{N}$	Flat	$48.07_{1.09}$	$49.28_{0.72}$	$86.38_{0.68}$	$40.26_{0.52}$	$47.59_{0.11}$	$78.01_{0.13}$
$\mathcal{M} \times \mathcal{N}$	MAG	$53.20_{2.13}$	$57.86_{1.07}$	$86.11_{0.12}$	$52.77_{0.02}$	$58.93_{0.11}$	$82.86_{0.13}$
$\mathcal{M} \times \mathcal{N}$	MulT	$54.13_{1.20}$	$58.94_{1.94}$	$87.41_{0.26}$	$52.73_{0.28}$	$58.89_{0.29}$	$83.01_{0.04}$
$\mathcal{M} \times \mathcal{N}$	UTDE	$56.45_{1.30}$	$60.23_{1.54}$	$88.80_{0.22}$	$52.10_{0.18}$	$58.90_{0.15}$	$83.36_{0.18}$
$\mathcal{M} \times \mathcal{N} \times \mathcal{K}$	**KIMP (Our)**	$\mathbf{58.14_{0.42}}$	$\mathbf{61.54_{0.44}}$	$\mathbf{90.08_{0.10}}$	$\mathbf{54.07_{0.15}}$	$\mathbf{59.57_{0.02}}$	$83.62_{0.06}$

4   Experiments

To demonstrate the effectiveness of our proposed methods, we conducted a series of rigorous experiments and ablation studies focusing on two vital medical tasks: 48-hour in-hospital mortality prediction (IHM) and 48-hour phenotype classification (PHE). These tasks hold significant importance within the clinical scenario [38,39], making them ideal benchmarks for evaluating the performance of our approach.

4.1   Experiment Setup

Dataset. MIMIC-III, an extensively used real-world public EHR dataset comprising patients admitted to Intensive Care Units (ICUs), encompasses numerical time series data as well as clinical notes [40]. We select the MISTS features and extract clinical notes following previous studies conducted by [14] and [41], respectively. For the preprocessing of the Unified Medical Language System (UMLS) [32], we follow the approach outlined in [42,43], which apply a named entity recognition, and linking tool ScispaCy [44] to pre-process the texts in the clinical notes to link entities in the texts to the UMLS knowledge base for entity disambiguation. Consistent with the methodology employed by [14], the data is split into training, validation, and testing sets. Furthermore, patients lacking any clinical notes preceding the prediction time are meticulously eliminated from the dataset to ensure data integrity and reliability. After preprocessing, the number of patients in the training, validation and testing sets for the IHM are 11181, 2473, 2488; and for the PHE, they are 22275, 4887, 4803, respectively. The number of domain knowledge entities is 5342, while the number of relationships is 404. The total number of knowledge triplets is 57250.

Baselines and Evaluation Metric. To examine the effectiveness of our KIMP method, we consider five baselines for medication prediction: mTAND [21], Flat [45], MAG [46], MulT [37], UTDE [22]. Specifically, the mTAND method is used for unimodal that encode the time information, the subsequent methods are multimodal method for the MISTS and clinical notes modals. We evaluated the performance of our proposed methods and baseline approaches following the previous works [14,41], including F-score, Area Under the Precision-Recall Curve (AUPR), and Area Under the Receiver Operating Characteristic Curve (AUROC). These metrics have been widely adopted in previous studies, ensuring consistency and comparability with the existing literature. By assessing the performance of our models on the tasks of 48-hour in-hospital mortality prediction (IHM) and 48-hour phenotype classification (PHE), we aim to provide a comprehensive evaluation of their effectiveness and robustness.

Parameter Setting. Here, we list the implementation details of KIMP. We set the hidden dimension as 128 and learning rate for text encoder of 2×10^{-5} and others of 4×10^{-4}. For the clinical notes encoder, we use the clinical long-former [47]. We use the Adam algorithm for gradient-based optimization [48]. We store the parameters that obtain the hightest F1 in the validation set, and use it to make predictions for testing samples. The memory slot number is 60. Our model is implemented by Pytorch 1.13 based on Python 3.8.18 and training on GeForce RTX 3090 GPUs.

5 Results

Main Results. In this section, we conduct a comprehensive comparison of results between our proposed method KIMP and their respective baselines in the MISTS, irregular clinical notes, and multimodal scenarios, respectively. As shown in Table 1, our propose model KIMP outperforms all baselines with the higher F1, AUPR and AUROC. From Table 1, we can also draw the following conclusions: For unimodal data, the performance of MISTS data yields significantly lower performance compared to the performance of clinical notes. This could be attributed to the clinical notes contain more comprehensive patient information, such as family history and personal details. Regarding multimodal approaches, we observe that different fusion strategies and sequence encoding methods have a substantial impact on the performance of medical prediction. Specifically, the Flat method directly encodes MISTS using LSTM without considering time information, and it concatenates MISTS and clinical notes without considering the interaction between modalities. As a result, the experimental results of the Flat method are not satisfactory. From the experimental results of the Flat method, we can conclude that the encoding methods for MISTS and clinical notes are crucial, and the fusion approach among modalities directly affects the performance of medical prediction. By comparing the methods that incorporate knowledge, we observe a significant improvement in experimental performance for both unimodal and multimodal data. For instance, in the IHM task using only clinical notes, the addition of knowledge leads to a 2.03% increase in F1 score and a 3.23% increase in AUPR. When compared to the best performing multimodal method, UTDE, our KIMP method achieves a 1.69% increase in F1 score, a 1.31% increase in AUPR, and a 1.28% increase in AUROC in the IHM task. In the PHE task, our KIMP method outperforms the UTDE method with a 1.97% increase in F1 score and a 0.67% increase in AUPR. Overall, by incorporating knowledge, our proposed KIMP method significantly enhances the performance of medical prediction.

Table 2. Ablation study on the effects of substituting different submodules in KIMP.

Model	modules			IHM			PHE		
	DKI	IKI	CMF	$F1_{std.}$	$AUPR_{std.}$	$AUROC_{std.}$	$F1_{std.}$	$AUPR_{std.}$	$AUROC_{std.}$
KIMP	✓	✓	✓	$\mathbf{58.14_{0.42}}$	$\mathbf{61.54_{0.44}}$	$\mathbf{90.08_{0.10}}$	$\mathbf{54.07_{0.15}}$	$\mathbf{59.57_{0.02}}$	$\mathbf{83.62_{0.06}}$
$KIMP_{w/o\ IKI,CMF}$	✓			$55.44_{0.74}$	$58.66_{0.68}$	$88.77_{0.43}$	$52.33_{1.02}$	$58.82_{0.05}$	$82.80_{0.22}$
$KIMP_{w/oCMF}$	✓	✓		$56.31_{0.90}$	$60.44_{0.16}$	$89.84_{0.19}$	$52.73_{0.07}$	$58.64_{0.04}$	$83.11_{0.02}$
$KIMP_{w/o\ IKI}$	✓		✓	$55.93_{0.51}$	$60.10_{0.69}$	$89.71_{0.07}$	$52.61_{0.08}$	$59.23_{0.23}$	$82.89_{0.11}$
$KIMP_{w/o\ DKI,CMF}$		✓		$54.88_{0.77}$	$58.48_{0.75}$	$88.61_{0.68}$	$52.36_{0.29}$	$58.73_{0.24}$	$82.60_{0.13}$
$KIMP_{w/o\ DKI}$		✓	✓	$56.67_{0.36}$	$59.76_{0.31}$	$89.61_{0.13}$	$52.59_{0.36}$	$59.22_{0.04}$	$83.13_{0.21}$
$KIMP_{w/o\ DKI,IKI}$			✓	$56.42_{0.34}$	$59.80_{0.28}$	$89.60_{0.01}$	$52.55_{0.26}$	$58.79_{0.01}$	$82.42_{0.11}$

Ablation Study. The success of KIMP is ascribed to the three modules we proposed (i.e., Domain Knowledge Injection (DKI), the In-memory Knowledge Injection (IKI), and the Cross Modal Fusion (CMF)). To verify the effectiveness of each module we proposed, we designed the ablation experiments, $KIMP_{w/oDKI}$ removes the domain knowledge injection module and directly utilizes the encoded representations of MISTS and clinical notes. $KIMP_{w/oIKI}$ removes the in-memory knowledge injection module. $KIMP_{w/oCMF}$ leverage concatenation operation to fuse the MISTS and clinical notes information for medical prediction. $KIMP_{w/oDKI,IKI}$ means only uses the MISTS and clinical notes representation into the cross modal fusion module. $KIMP_{w/oDKI,CMF}$ and $KIMP_{w/oIKI,CMF}$ represent the removal of the domain knowledge injection and in-memory knowledge injection modules, respectively, while also excluding the cross modal fusion module. Table 2 shows the results for the different variants of KIMP. As expected, when randomly removing the three modules we proposed. The performance brought a significant deterioration to the complete KIMP model. Overall, our KIMP method outperforms all the variants, which demonstrates the effectiveness of each module proposed and signifies that each component is an indispensable part of KIMP (Fig. 3).

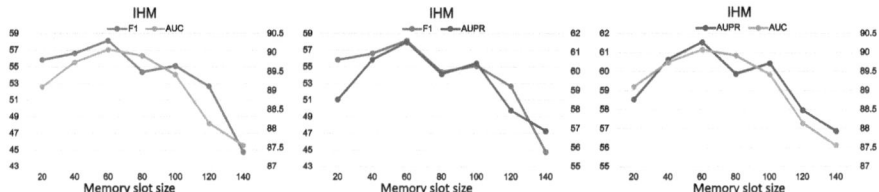

Fig. 3. Performance of KIMP method along with different memory slot sizes.

Discussion on the Size of Sample-Based memory slot. To investigate the impact of the size of the sample-based memory slot on experimental results, we conducted separate discussions on the performance differences of KIMP in the IHM task in terms of F1 score, AUPR, and AUC, under different memory slot sizes. As shown in Fig. 3, we can observe that we can observe that as the memory slot size increases up to 60, our proposed method KIMP exhibits a continuous improvement in performance. However, after surpassing a memory slot size of 60, KIMP's performance notably starts to decline. Therefore, we conclude that setting the memory slot size for the sample-based memory is not a case of "bigger is better." This observation suggests that our model typically only needs to retain the most crucial sample-based information.

6   Conclusion

In this paper, we propose a knowledge injected multimodal irregular EHR model for medication prediction. We first tack irregular MISTS and clinical notes data via a time attention mechanism separately, and effectively integrate static domain knowledge by a domain knowledge injection module. We also leverage dynamic memory mechanisms to integrate the knowledge stored in samples and inject it into the patient's representation. Furthermore, we introduce a cross modal fusion module to fuse the multimodal irregular EHR effectively. We hope that our work will inspire further exploration in addressing the challenges of handling multimodal EHR, as well as utilizing domain knowledge in medical prediction tasks.

Acknowledgments. This work was supported by the Hong Kong Innovation and Technology Fund (Project No. MHP/002/22), HKUST (Project No. FS111) and Research Grants Council of the Hong Kong (No. R6003-22 and T45-401/22-N).

References

1. Tisherman, S.A., Stein, D.M.: ICU management of trauma patients. Crit. Care Med. **46**(12), 1991–1997 (2018)
2. Alberti, C., et al.: Epidemiology of sepsis and infection in ICU patients from an international multicentre cohort study. Intensive Care Med. **28**, 108–121 (2002)
3. Afessa, B., Gajic, O., Keegan, M.T.: Severity of illness and organ failure assessment in adult intensive care units. Crit. Care Clin. **23**(3), 639–658 (2007)
4. Cullen, D.J., Sweitzer, B.J., Bates, D.W., Burdick, E., Edmondson, A., Leape, L.L.: Preventable adverse drug events in hospitalized patients: a comparative study of intensive care and general care units. Crit. Care Med. **25**(8), 1289–1297 (1997)
5. Otero-López, M.J., Alonso-Hernández, P., Maderuelo-Fernández, J.A., Garrido-Corro, B., Domínguez-Gil, A., Sánchez-Rodríguez, A.: Preventable adverse drug events in hospitalized patients. Med. Clin. **126**(3), 81–87 (2006)
6. Adler-Milstein, J., et al.: Electronic health record adoption in us hospitals: progress continues, but challenges persist. Health Aff. **34**(12), 2174–2180 (2015)
7. Atasoy, H., Greenwood, B.N., McCullough, J.S.: The digitization of patient care: a review of the effects of electronic health records on health care quality and utilization. Annu. Rev. Public Health **40**, 487–500 (2019)
8. Rajkomar, A., et al.: Scalable and accurate deep learning with electronic health records. NPJ Digit. Med. **1**(1), 1–10 (2018)
9. Xiao, C., Choi, E., Sun, J.: Opportunities and challenges in developing deep learning models using electronic health records data: a systematic review. J. Am. Med. Inform. Assoc. **25**(10), 1419–1428 (2018)
10. Ahmed, S.F., et al.: Deep learning modelling techniques: current progress, applications, advantages, and challenges. Artif. Intell. Rev. **56**(11), 13521–13617 (2023)
11. Shickel, B., Tighe, P.J., Bihorac, A., Rashidi, P.: Deep EHR: a survey of recent advances in deep learning techniques for electronic health record (EHR) analysis. IEEE J. Biomed. Health Inform. **22**(5), 1589–1604 (2017)
12. Xie, F., Chakraborty, B., Ong, M.E.H., Goldstein, B.A., Liu, N., et al.: Autoscore: a machine learning-based automatic clinical score generator and its application to mortality prediction using electronic health records. JMIR Med. Inform. **8**(10), e21798 (2020)
13. Khojandi, A., Tansakul, V., Li, X., Koszalinski, R.S., Paiva, W.: Prediction of sepsis and in-hospital mortality using electronic health records. Methods Inf. Med. **57**(04), 185–193 (2018)
14. Harutyunyan, H., Khachatrian, H., Kale, D.C., Ver Steeg, G., Galstyan, A.: Multitask learning and benchmarking with clinical time series data. Sci. Data **6**(1), 96 (2019)
15. Bhoi, S., Lee, M.L., Hsu, W., Fang, H.S.A., Tan, N.C.: Personalizing medication recommendation with a graph-based approach. ACM Trans. Inf. Syst. (TOIS) **40**(3), 1–23 (2021)
16. Liu, S., et al.: Shape: a sample-adaptive hierarchical prediction network for medication recommendation. IEEE J. Biomed. Health Inform. (2023)
17. Zerveas, G., Jayaraman, S., Patel, D., Bhamidipaty, A., Eickhoff, C.: A transformer-based framework for multivariate time series representation learning. In: Proceedings of the 27th ACM SIGKDD Conference on Knowledge Discovery & Data Mining, pp. 2114–2124 (2021)
18. Che, Z., Purushotham, S., Cho, K., Sontag, D., Liu, Y.: Recurrent neural networks for multivariate time series with missing values. Sci. Rep. **8**(1), 6085 (2018)

19. Zhang, X., Zeman, M., Tsiligkaridis, T., Zitnik, M.: Graph-guided network for irregularly sampled multivariate time series. arXiv preprint arXiv:2110.05357 (2021)
20. Shukla, S.N., Marlin, B.M.: Interpolation-prediction networks for irregularly sampled time series. arXiv preprint arXiv:1909.07782 (2019)
21. Shukla, S.N., Marlin, B.M.: Multi-time attention networks for irregularly sampled time series. arXiv preprint arXiv:2101.10318 (2021)
22. Zhang, X., Li, S., Chen, Z., Yan, X., Petzold, L.R.: Improving medical predictions by irregular multimodal electronic health records modeling. In: International Conference on Machine Learning, pp. 41300–41313. PMLR (2023)
23. Golmaei, S.N., Luo, X.: Deepnote-GNN: predicting hospital readmission using clinical notes and patient network. In: Proceedings of the 12th ACM Conference on Bioinformatics, Computational Biology, and Health Informatics, pp. 1–9 (2021)
24. Mahbub, M., et al.: Unstructured clinical notes within the 24 hours since admission predict short, mid & long-term mortality in adult ICU patients. PLoS ONE **17**(1), e0262182 (2022)
25. Huang, K., Altosaar, J., Ranganath, R.: Clinicalbert: modeling clinical notes and predicting hospital readmission. arXiv preprint arXiv:1904.05342 (2019)
26. Gu, Y., et al.: Domain-specific language model pretraining for biomedical natural language processing. ACM Trans. Comput. Healthc. (HEALTH) **3**(1), 1–23 (2021)
27. Zhang, D., Thadajarassiri, J., Sen, C., Rundensteiner, E.: Time-aware transformer-based network for clinical notes series prediction. In: Machine Learning for Healthcare Conference, pp. 566–588. PMLR (2020)
28. Ma, F., You, Q., Xiao, H., Chitta, R., Zhou, J., Gao, J.: Kame: knowledge-based attention model for diagnosis prediction in healthcare. In: Proceedings of the 27th ACM International Conference on Information and Knowledge Management, pp. 743–752 (2018)
29. Choi, E., Bahadori, M.T., Song, L., Stewart, W.F., Sun, J.: Gram: graph-based attention model for healthcare representation learning. In: Proceedings of the 23rd ACM SIGKDD International Conference on Knowledge Discovery and Data Mining, pp. 787–795 (2017)
30. Mullenbach, J., Wiegreffe, S., Duke, J., Sun, J., Eisenstein, J.: Explainable prediction of medical codes from clinical text. arXiv preprint arXiv:1802.05695 (2018)
31. Niu, S., Yin, Q., Song, Y., Guo, Y., Yang, X.: Label dependent attention model for disease risk prediction using multimodal electronic health records. In: 2021 IEEE International Conference on Data Mining (ICDM), pp. 449–458. IEEE (2021)
32. Bodenreider, O.: The unified medical language system (UMLS): integrating biomedical terminology. Nucleic Acids Res. **32**(suppl_1), D267–D270 (2004)
33. Kazemi, S.M., et al.: Time2vec: learning a vector representation of time. arXiv preprint arXiv:1907.05321 (2019)
34. Vaswani, A., et al.: Attention is all you need. In: Advances in Neural Information Processing Systems, vol. 30 (2017)
35. Bordes, A., Usunier, N., Garcia-Duran, A., Weston, J., Yakhnenko, O.: Translating embeddings for modeling multi-relational data. In: Advances in Neural Information Processing Systems, vol. 26 (2013)
36. Veličković, P., Cucurull, G., Casanova, A., Romero, A., Lio, P., Bengio, Y.: Graph attention networks. arXiv preprint arXiv:1710.10903 (2017)
37. Tsai, Y.-H.H., Bai, S., Liang, P.P., Kolter, J.Z., Morency, L.-P., Salakhutdinov, R.: Multimodal transformer for unaligned multimodal language sequences. In: Proceedings of the Conference. Association for Computational Linguistics Meeting, vol. 2019, p. 6558. NIH Public Access (2019)

38. Choi, E., Bahadori, M.T, Schuetz, A., Stewart, W.F., Sun, J.: Doctor AI: predicting clinical events via recurrent neural networks. In: Machine Learning for Healthcare Conference, pp. 301–318. PMLR (2016)

39. Gupta, P., Malhotra, P., Narwariya, J., Vig, L., Shroff, G.: Transfer learning for clinical time series analysis using deep neural networks. J. Healthc. Inform. Res. **4**(2), 112–137 (2020)

40. Johnson, A.E., et al.: MIMIC-III, a freely accessible critical care database. Sci. Data **3**(1), 1–9 (2016)

41. Khadanga, S., Aggarwal, K., Joty, S., Srivastava, J.: Using clinical notes with time series data for ICU management. arXiv preprint arXiv:1909.09702 (2019)

42. Liu, S., Wang, X., Zhao, X., Chen, H.: Medication recommendation via domain knowledge informed deep learning. arXiv preprint arXiv:2305.19604 (2023)

43. Chen, Z., Li, G., Wan, X.: Align, reason and learn: Enhancing medical vision-and-language pre-training with knowledge. In: Proceedings of the 30th ACM International Conference on Multimedia, pp. 5152–5161 (2022)

44. Neumann, M., King, D., Beltagy, I., Ammar, W.: Scispacy: fast and robust models for biomedical natural language processing. arXiv preprint arXiv:1902.07669 (2019)

45. Deznabi, I., Iyyer, M., Fiterau, M.: Predicting in-hospital mortality by combining clinical notes with time-series data. Find. Assoc. Comput. Linguist. ACL-IJCNLP **2021**, 4026–4031 (2021)

46. Yang, B., Wu, L.: How to leverage the multimodal EHR data for better medical prediction? In: Proceedings of the 2021 Conference on Empirical Methods in Natural Language Processing, pp. 4029–4038 (2021)

47. Li, Y., Wehbe, R.M., Ahmad, F.S., Wang, H., Luo, Y.: A comparative study of pretrained language models for long clinical text. J. Am. Med. Inform. Assoc. **30**(2), 340–347 (2023)

48. Diederik, P.K.: Adam: a method for stochastic optimization (2014)

FusionINN: Decomposable Image Fusion for Brain Tumor Monitoring

Nishant Kumar[1]($\boxtimes$), Ziyan Tao[1], Jaikirat Singh[1], Yang Li[2], Peiwen Sun[3], Binghui Zhao[3], and Stefan Gumhold[1]

[1] Chair of Computer Graphics and Visualization, Faculty of Computer Science, Technische Universität Dresden, Dresden, Germany
nishant.kumar@tu-dresden.de
[2] School of Computer Science and Engineering, Shandong University of Science and Technology, Qingdao, China
[3] Department of Radiology, Shanghai Tenth People's Hospital, Tongji University Medical School, Shanghai, China

Abstract. Image fusion typically employs non-invertible neural networks to merge multiple source images into a single fused image. However, for clinical experts, solely relying on fused images may be insufficient for making diagnostic decisions, as the fusion mechanism blends features from source images, thereby making it difficult to interpret the underlying tumor pathology. We introduce FusionINN, a novel decomposable image fusion framework, capable of efficiently generating fused images and also decomposing them back to the source images. FusionINN is designed to be bijective by including a latent image alongside the fused image, while ensuring minimal transfer of information from the source images to the latent representation. To the best of our knowledge, we are the first to investigate the decomposability of fused images, which is particularly crucial for life-sensitive applications such as medical image fusion compared to other tasks like multi-focus or multi-exposure image fusion. Our extensive experimentation validates FusionINN over existing discriminative and generative fusion methods, both subjectively and objectively. Moreover, compared to a recent denoising diffusion-based fusion model, our approach offers faster and qualitatively better fusion results. The source code of the FusionINN framework is available at: https://github.com/nish03/FusionINN.

Keywords: Medical Image Fusion · Image Decomposition · Generative Model · Normalizing Flows · Invertible Neural Networks (INNs)

1 Introduction

Magnetic Resonance Imaging (MRI) techniques, such as Diffusion-weighted imaging with Apparent Diffusion Coefficient (DWI-ADC) and T2-weighted Fluid Attenuated Inversion Recovery (T2-Flair), offer invaluable insights into the intricate pathology of tumors. A high-intensity signal on the T2-Flair image provides

H. Chen et al. (Eds.): TAI4H 2024, LNCS 14812, pp. 40–51, 2024.
https://doi.org/10.1007/978-3-031-67751-9_4

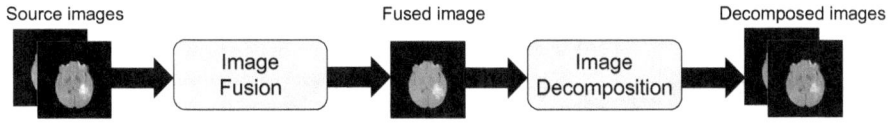

Fig. 1. An illustration of the task of image fusion and decomposition.

anatomical information about the presence of tumor and its boundary [1]. In contrast, DWI-ADC assists in revealing the tumor category, as a high-intensity signal indicates the existence of liquid components, i.e., necrotic tumor tissues and a low-intensity signal suggests the presence of solid components, i.e., enhancing tumor tissues [2]. Clinicians commonly utilize such image modalities postoperatively to detect any residual necrotic tumor tissues and assess the potential for its recurrence by locating enhancing tumor tissues. Fused images can aid in the visualization of the clinical features from multiple sources. However, merging grayscale values can obscure salient features, thereby complicating clinical interpretation of the fused image. To address this problem, we introduce the extended fusion task illustrated in Fig. 1, which demands decomposability of the fused image into the source images.

Prior works in image fusion leverage deep learning algorithms via discriminative training [3–8,11,29,32] or generative modeling using generative adversarial networks (GANs) [9]. However, the network architecture of such image fusion approaches is not invertible. As a result, they have not been utilized for decomposing fused images. Recently, a pre-trained Denoising Diffusion based image fusion model [10] has been proposed, that conditions each of the denoising diffusion steps on source images. In principle, diffusion models allow stable training dynamics, while not suffering from mode collapse. However, the decomposability of the fused images is also not explored in [10], possibly because the pre-trained UNet [18] model used to perform the denoising steps is not invertible. Additionally, diffusion models perform slow sequential sampling through multiple denoising steps to obtain the fusion output, due to which a real-time inference scheme is impractical.

We present normalizing flows as the generative model for medical image fusion and capitalize on their inherent invertibility to facilitate the decomposability of the fusion process. The flow demonstrates efficient sampling capabilities and stability during training through the use of invertible transformations, which are beneficial for computer vision tasks [24,27]. Previous attempts utilizing invertible neural networks (INNs) for image fusion [12–16] have predominantly integrated INNs only as a sub-module within a multi-step pipeline, preventing the invertibility of the end-to-end fusion procedure. Notably, no prior studies have explored solving both the tasks of image fusion and decomposition through an end-to-end INN model. The primary contributions of this work are as follows:

– We introduce a first-of-its-kind image fusion framework, FusionINN, that harnesses invertible normalizing flow for bidirectional training. FusionINN not

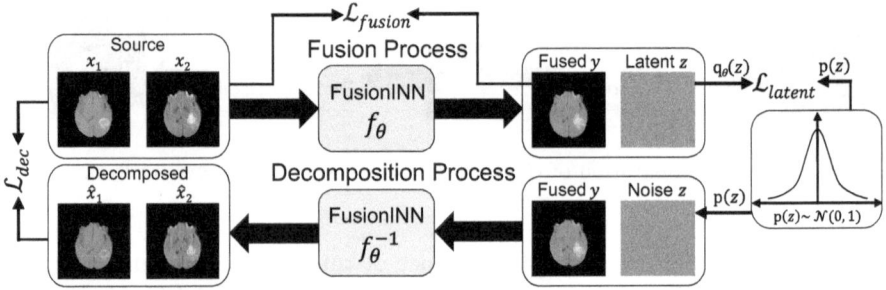

Fig. 2. An overview of the FusionINN framework.

only generates a fused image but can also decompose it into constituent source images, thus enhancing the interpretability for clinical practitioners.

- We present an extensive evaluation study that shows state-of-the-art results of FusionINN with common fusion metrics, alongside its additional capability to decompose the fused images.
- We also illustrate the effectiveness of FusionINN in fusing and decomposing images from clinical modalities that were not encountered during training.

2   Method

The objective under decomposable image fusion, as depicted in Fig. 1, is to generate a fused image that closely resembles the source images and can be decomposed back into those source images without additional information.

2.1   INN-Based Decomposable Image Fusion

The FusionINN framework for decomposable image fusion is shown in Fig. 2. In the forward fusion process, the FusionINN transforms the two source images $x_1 \in \mathbb{R}^n$ and $x_2 \in \mathbb{R}^n$ to a fused image $y \in \mathbb{R}^n$ and a latent image $z \in \mathbb{R}^n$ using the normalizing flow network f with parameters θ such that $[y, z] = f_\theta(x_1, x_2)$, where n is the number of pixels in the four equal resolution images. Consequently, the dimensionality of $[y, z]$ matches $[x_1, x_2]$ with $f_\theta, f_\theta^{-1} : \mathbb{R}^{2n} \leftrightarrow \mathbb{R}^{2n}$. Unlike GANs, which adversarially train two separate neural networks, normalizing flow requires training only a single network. This simplifies the training and makes it more stable, as there is no adversarial training dynamics. We introduce the latent image z to ensure the decomposability of the fused image, as the reverse mapping from a fused image to two source images is ill-posed. As the latent image is unknown for the decomposition task, we aim to capture as few source image features as possible in it. Therefore, we define the latent image z to follow a multivariate normal distribution, such that $z \sim p(z) = \mathcal{N}(z; 0, I)$. However, other design choices, such as a constant image z, are also feasible. Finally, the decomposition process utilizes a newly sampled latent image z along with the fused image y through the reverse direction of FusionINN i.e., f_θ^{-1} to produce the decomposed images $\hat{x}_1$ and $\hat{x}_2$, such that $[\hat{x}_1, \hat{x}_2] = f_\theta^{-1}([y, z])$. The inverse function f_θ^{-1} should learn to decompose the fused image y, independently from the latent image z, while ensuring that the decomposed images $\hat{x}_1$ and $\hat{x}_2$ closely resemble the source images x_1 and x_2.

2.2   INN Architecture

The FusionINN as a normalizing flow network f_θ consists of k invertible coupling blocks stacked together such that $f = f_k \circ ...f_j \circ ...f_1$ with $[\hat{x}_1, \hat{x}_2] = f_\theta^{-1}(y, z)$ and $[y, z] = f_\theta(x_1, x_2)$. In [21], the coupling blocks consist of learnable affine functions, namely scaling (s_1 and s_2) and translation (t_1 and t_2). We define these functions as convolutional neural networks (CNNs) with two convolutional layers, each followed by a ReLU activation. The input to an arbitrary j^{th} coupling block is first split into two parts u_1^j and u_2^j, which are transformed by s_1, t_1 and s_2, t_2 networks that share the learnable parameters. The output of the jth coupling block is the concatenation of the resulting parts v_1^j and v_2^j given as:

$$v_1^j = u_1^j \odot \exp\left(s_2(u_2^j)\right) + t_2(u_2^j), \quad v_2^j = u_2^j \odot \exp\left(s_1(v_1^j)\right) + t_1(v_1^j) \quad (1)$$

where $\odot$ is the element-wise multiplication, and the exponential term ensures non-zero coefficients. By construction, such a transformation is invertible, and u_1^j, u_2^j can be recovered from v_1^j, v_2^j (see [31]). Between each coupling block, we implement a random permutation operation to reorganize the two channels obtained from the output of the previous block. This permutation is applied only once and remains fixed during the training of FusionINN's learnable parameters θ. Furthermore, following the channel permutation, we utilize an invertible downsampling operator [35] to reduce the spatial resolution of the input channels without losing any information. For example, when $k = 3$, an invertible downsampling operation precedes the second coupling block, and an invertible upsampling operation is applied before the third coupling block to maintain the resolution of the final output of the normalizing flow network f_θ. This operation enables the network to increase its receptive field and effectively capture features at multiple scales. We also apply a sigmoid function as the final layer of the network to obtain the normalized fused image y.

2.3   Unsupervised Learning

The learning scheme of our FusionINN framework, depicted in Fig. 2, operates without a predefined fusion groundtruth. Therefore, we approach the fusion task as a fully unsupervised problem, utilizing the fusion loss $\mathcal{L}_{fusion}$. This loss function allows FusionINN to optimize the fused image without explicit supervision, learning directly from the source images. Additionally, FusionINN learns to shape the latent image z to conform to a standard normal distribution through the $\mathcal{L}_{latent}$ loss, which minimizes information transfer from source images to the

latent image. We also define a decomposition loss as $\mathcal{L}_{dec}$, which aids in estimating the source images from the fused image. With these loss functions, our
FusionINN framework not only achieves superior fusion results but also facilitates image decomposition.

Fusion Loss: To learn the fused image y from the flow network f_θ in an unsupervised manner, we follow [3] and leverage the metric Structural Similarity
Index (Q_{SSIM}) [22] as the differentiable loss function to maximize the similarity
between the source and the fused images. The loss function is formulated as:

$$\mathcal{L}_{SSIM} = \{1 - Q_{SSIM}(x_1, y)\} + \{1 - Q_{SSIM}(x_2, y)\} \qquad (2)$$

The sub-loss terms in $\mathcal{L}_{SSIM}$ are subtracted from 1 to satisfy the loss
minimization objective, as Q_{SSIM} computes the similarity between the two
images. However, while Q_{SSIM} is effective in preserving the structure and the
contrast of an image, it can alter the brightness and make the image appear
duller, as discussed in [28]. To address this, we use the squared ℓ_2 loss in addition to the Q_{SSIM} metric to better preserve the luminance of the fused image,
as squared ℓ_2 loss directly penalizes differences in pixel intensities. Finally, given
the weightage parameter as λ, the $\mathcal{L}_{\ell_2}$ and $\mathcal{L}_{fusion}$ losses are expressed as:

$$\mathcal{L}_{\ell_2} = ||y - x_1||_2^2 + ||y - x_2||_2^2, \quad \mathcal{L}_{fusion} = \{\lambda \mathcal{L}_{SSIM} + (1 - \lambda)\mathcal{L}_{\ell_2}\} \qquad (3)$$

Latent Loss: We model the distribution of the latent image z with a multivariate Gaussian $p(z) = \mathcal{N}(z; 0, I)$. We utilize Maximum Mean Discrepancy
(MMD) [30] as the loss function to quantify the difference between the probability distribution $p(z)$ and the distribution $q_\theta(z)$ of the latent image z generated
by the forward process of the FusionINN model, f_θ. Consequently, the latent
loss $\mathcal{L}_{latent}$ is defined as $\mathcal{L}_{latent} = \text{MMD}(q_\theta(z) \| p(z))$. This enables the learned
distribution $q_\theta(z)$ to be approximated as the standard normal distribution $p(z)$
after minimization of the $\mathcal{L}_{latent}$ loss.

Decomposition Loss: We define the decomposition loss $\mathcal{L}_{dec}$ in the reverse
direction of f_θ i.e. f_θ^{-1} to decompose the fused image y back to the source
images, using a newly sampled latent image z. We implement the $\mathcal{L}_{dec}$ loss as
the combination of the $\mathcal{L}_{dec}^{SSIM}$ and $\mathcal{L}_{dec}^{\ell_2}$ losses, which are weighted using the
meta-parameter λ, similar to the $\mathcal{L}_{fusion}$ loss. The $\mathcal{L}_{dec}^{SSIM}$ loss employs the
Q_{SSIM} metric, while $\mathcal{L}_{dec}^{\ell_2}$ computes the squared ℓ_2-loss to measure the dissimi-
larity between the decomposed and source images. Hence, given the decomposed
images $\hat{x}_1$ and $\hat{x}_2$, the losses $\mathcal{L}_{dec}^{SSIM}$, $\mathcal{L}_{dec}^{\ell_2}$ and $\mathcal{L}_{dec}$ are computed as:

$$\mathcal{L}_{dec}^{SSIM} = \{1 - Q_{SSIM}(x_1, \hat{x}_1)\} + \{1 - Q_{SSIM}(x_2, \hat{x}_2)\}$$
$$\mathcal{L}_{dec}^{\ell_2} = ||\hat{x}_1 - x_1||_2^2 + ||\hat{x}_2 - x_2||_2^2, \quad \mathcal{L}_{dec} = \lambda \mathcal{L}_{dec}^{SSIM} + (1 - \lambda)\mathcal{L}_{dec}^{\ell_2} \qquad (4)$$

Total Loss: In the forward process, the FusionINN optimizes the mapping
$[y, z] = f_\theta(x_1, x_2)$ using $\mathcal{L}_{fusion}$ and $\mathcal{L}_{latent}$ losses. Additionally, FusionINN's

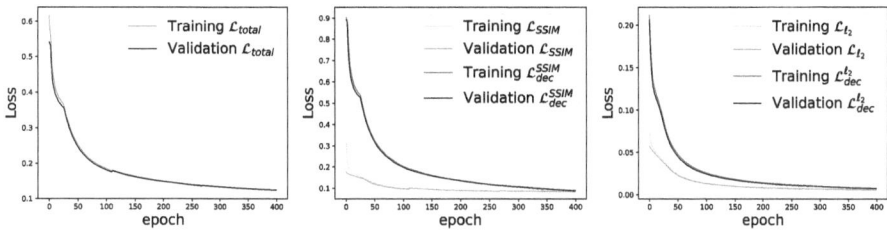

Fig. 3. Loss curves for a FusionINN training instance with $k = 4$, $\lambda = 0.9$ and $\alpha = 0.5$.

invertibility guarantees that the latent image generated from the forward process can precisely reproduce the source images. However, during the reverse process, we sample a new latent image z from the normal distribution $p(z)$ to maximize the decomposition accuracy, independent from any specific choice of z. The resampled latent image z, together with the fused image y is used to perform the reverse process by optimizing the mapping $[\hat{x}_1, \hat{x}_2] = f_\theta^{-1}(y, z)$ via the $\mathcal{L}_{dec}$ loss. Finally, we weight the forward losses i.e., $\mathcal{L}_{fusion}$ and $\mathcal{L}_{latent}$ as well as the decomposition loss i.e., $\mathcal{L}_{dec}$ using the parameter α and formulate the total loss function $\mathcal{L}_{total}$ as follows:

$$\mathcal{L}_{total} = \{\alpha(\mathcal{L}_{fusion} + \mathcal{L}_{latent}) + (1 - \alpha)\mathcal{L}_{dec}\} \tag{5}$$

Training Procedure: We learn the FusionINN's parameters θ by iteratively optimizing them to minimize the total loss function, $\mathcal{L}_{total}$. This involves computing the gradients of $\mathcal{L}_{total}$ with respect to each parameter using backpropagation and updating the parameters using Adam optimization [36] with a learning rate of 3×10^{-4}. The training is performed over 400 epochs with a batch size of 64. We also utilize a learning rate scheduler to reduce the learning rate by a factor of 0.95 if the validation loss does not improve for eight epochs, preventing the model from getting stuck in local minima and ensuring smoother convergence. The loss curves for $\mathcal{L}_{total}$ and the sub-losses $\mathcal{L}_{SSIM}$, $\mathcal{L}_{dec}^{SSIM}$, $\mathcal{L}_{\ell_2}$ and $\mathcal{L}_{dec}^{\ell_2}$ at each training epoch are illustrated in Fig. 3.

3   Results and Discussion

Data Description: We use the publicly available BraTS-2018 brain imaging dataset [34] to prepare our training and validation data. We extract post-contrast T1-weighted (T1-Gd) and T2-Flair as the two source images, acquired with different clinical protocols and different scanners from multiple medical institutions. The data has been pre-processed, i.e., co-registered to the same anatomical template, interpolated to the same resolution and skull-stripped [34]. We only extract those images from the dataset where the clinical annotation comprises of the necrotic core, non-enhancing tumor, and the peritumoral edema. This results

Table 1. Comparison of the fusion performance of the evaluated models on the validation set of our pre-processed BraTS-2018 images [34]. The results from each model show averaged scores from five metrics after comparing the fused images with the source image pairs. For each metric, the best-performing model is highlighted in bold.

Model Type	Model Name	Q_{SSIM} ↑	Q_{FMI} ↑	Q_{NCIE} ↑	Q_{XY} ↑	Q_P ↑
Discriminative	DeepFuse [3]	0.927	0.791	0.806	0.449	0.766
(Equal Dimension)	FunFuseAn [5]	0.930	0.845	0.806	0.481	0.781
Discriminative	Half-UNet [17]	0.933	0.850	0.805	0.464	0.774
(Dimension Reduction)	UNet [18]	0.934	0.835	0.805	0.420	0.711
	UNet++ [19]	**0.937**	0.849	0.805	0.433	0.739
	UNet3+ [20]	0.937	0.849	0.805	0.434	0.742
Generative	DDFM [10]	0.921	**0.861**	0.806	0.472	0.702
	FusionINN (Ours)	0.927	0.835	**0.806**	**0.493**	**0.783**

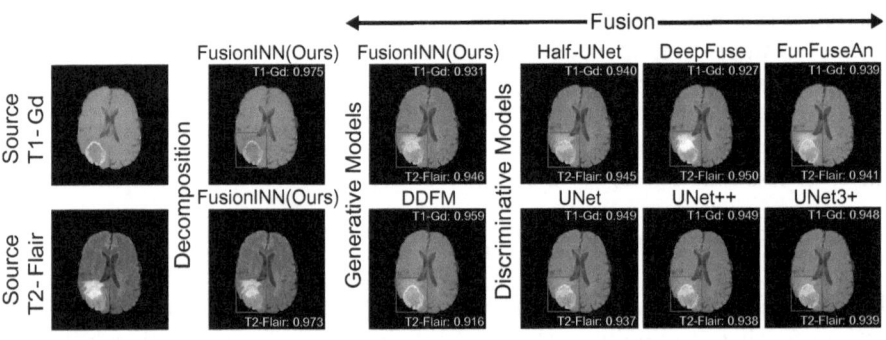

Fig. 4. Fusion results obtained from the evaluated models on a sample validation image pair. The Q_{SSIM} scores for individual modalities are shown in the fused images.

in 9653 image pairs of T1-Gd and T2-Flair modalities. We randomly assign 8500 image pairs as training and 1153 image pairs as the validation set.

Competitive Methods and Evaluation Metrics: We assess FusionINN's performance by comparing it with other fusion methods, namely DeepFuse [3] and FunFuseAn [5]. We also repurpose popular image segmentation models namely Half-UNet [17], UNet [18], UNet++ [19], and UNet3+ [20] for the image fusion task. Each of these models involve discriminative modeling and are trained on the fusion loss, i.e., $\mathcal{L}_{fusion}$ (Eq. 3). These models are non-invertible and can only be used to estimate fused images. We maintain a common benchmark of meta-parameters during training of these models. Furthermore, we employ the pre-trained Denoising Diffusion-based Fusion model (DDFM) [10] as a generative method to evaluate its performance on our validation images. We utilize five quantitative metrics specifically designed for assessing the image fusion quality. The metrics are Feature Mutual Information (Q_{FMI}) [33], Structural Similarity Index (Q_{SSIM}) [22], Non-linear Correlation Information Entropy

(Q_{NCIE}) [26], and by Petrovic et al. (Q_{XY}) [23], and Piella et al. (Q_P) [25]. The metric Q_{XY} use gradient representation of the source images to quantify the information or feature transfer to the fused images. On the other hand, Q_P weights the structural similarity scores based on local saliencies of the two source images.

Fusion and Decomposition Performance: Table 1 presents the quantitative fusion results of the evaluated models across various fusion metrics after averaging over the validation images. Our FusionINN model demonstrates either comparable or superior fusion performance with respect to all other evaluated models across metrics such as Q_{NCIE}, Q_{XY}, and Q_P. Notably, FusionINN also exhibits competitive results on Q_{SSIM} metric. The Fig. 4 shows qualitative fusion results using a sample image pair from the validation set. The fusion results from the FusionINN model is competitive with other methods, and its decomposition results closely resemble the source images. Despite UNet-based methods exhibiting comparable Q_{SSIM} scores, FusionINN excels in preserving the high-intensity features from the T2-Flair image within the fused output.

Table 2. The effect of coupling blocks k, latent image z, and parameters λ and α on the fusion and decomposition performance is examined. The results are obtained from a single training run, using different initial random seeds for each combination of meta-parameters. These results are averaged Q_{SSIM} scores over the validation images, with $Q_{SSIM}(x, y)$ for fusion and $Q_{SSIM}(x, \hat{x})$ for decomposition. When studying α, λ, and k, we maintain $z \sim \mathcal{N}(0, I)$. Additionally, we fix $k = 3$ when analyzing the impact of different types of latent image z on the fusion and decomposition results.

Weight (α)	Fusion		Decomposition		Weight (λ)	Fusion		Decomposition	
($k = 3, \lambda = 0.8$)	T1-Gd	T2-Flair	T1-Gd	T2-Flair	($k = 3, \alpha = 0.5$)	T1-Gd	T2-Flair	T1-Gd	T2-Flair
0.2	0.903	0.929	0.930	0.930	0.8	0.921	**0.933**	**0.976**	0.972
0.5	0.921	**0.933**	**0.976**	**0.972**	0.9	0.925	0.929	0.974	**0.974**
0.8	0.926	0.933	0.927	0.920	0.99	0.915	0.928	0.969	0.974
1.0	**0.948**	0.898	0.033	0.004	0.999	**0.935**	0.923	0.937	0.920

Blocks (k)	Fusion		Decomposition		Latent (z)	Fusion		Decomposition	
($\alpha = 0.5, \lambda = 0.8$)	T1-Gd	T2-Flair	T1-Gd	T2-Flair	($\alpha = 0.5, \lambda = 0.8$)	T1-Gd	T2-Flair	T1-Gd	T2-Flair
1	0.914	**0.943**	0.151	0.093	0	0.918	0.930	0.932	0.928
2	**0.935**	0.923	0.945	0.928	$\mathcal{N}(0, I)$	0.921	**0.933**	**0.976**	**0.972**
3	0.921	0.933	**0.976**	**0.972**	$\mathcal{U}[0, I)$	**0.924**	0.925	0.958	0.954
4	0.923	0.936	0.937	0.939	1	0.916	0.929	0.967	0.969

Ablation Studies: Table 2 demonstrates the impact of various parameters on FusionINN's fusion and decomposition performance. The results in the upper left portion of Table 2 indicate that three coupling blocks with $\lambda = 0.8$ and $\alpha = 0.5$ produce competitive results in terms of Q_{SSIM} scores. Furthermore, increasing α enhances image fusion performance with respect to at least one source modality. This can be attributed to a higher weightage given to the $\mathcal{L}_{fusion}$ loss in

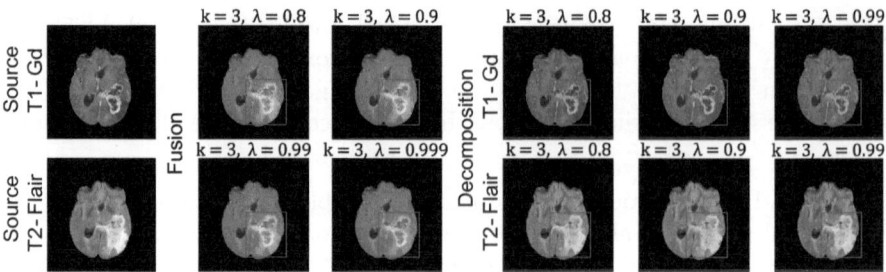

Fig. 5. FusionINN results at $\alpha = 0.5$, $z \sim \mathcal{N}(0, I)$ and k as number of coupling blocks.

the optimization process. We also explored different latent priors for z, including learning zeros, ones, and a uniform distribution $\mathcal{U}[0, 1)$. The results in the bottom right portion of Table 2 show that, on average, the fusion performance is similar under each type of latent prior for z. This indicates that the latent image z does not influence the construction and quality of the fused images. Furthermore, interpreting the decomposition performance, it can be argued that, on average, a constant image z with only ones in its pixel values performs almost as good as a latent image z defined with random noise. In Fig. 5, the qualitative fusion and decomposition results portray that both $\lambda = 0.8$ and 0.9 provide a good compensation of Q_{SSIM} via squared ℓ_2-loss, resulting in superior visual quality of the images.

Clinical Translation: In this study, we aimed to evaluate the robustness of the FusionINN model for practical clinical usage. To achieve this, we assessed the model's performance on entirely new and clinically relevant test modalities that were not included in the training data. Figure 6 illustrates clinically acquired image pairs from DWI-ADC and T2-Flair modalities, showing post-operative tumor regions of two patients following brain surgery. The medical practitioners sought both fused and decomposed images of the test image pairs shown in Fig. 6 to better evaluate the model's efficacy in aiding prognosis. Specifically, the model was expected to produce images that clearly delineate features in T2-Flair indicative of the tumor's anatomical boundary, while also preserving high- and low-intensity DWI-ADC features related to residual necrotic and enhancing tumor tissues. Note that the FusionINN model was trained on image pairs of T1-Gd and T2-Flair modalities. The results shown in Fig. 6 demonstrate that the model preserves salient features from both modalities in its decomposed images and effectively combines source features into the fused image. These findings highlight the efficacy and generalization capability of the model to accurately construct fused and decomposed images, even for unseen test images from new image modalities. Therefore, the clinically robust results obtained from the FusionINN model should assist clinicians in making better diagnostic decisions.

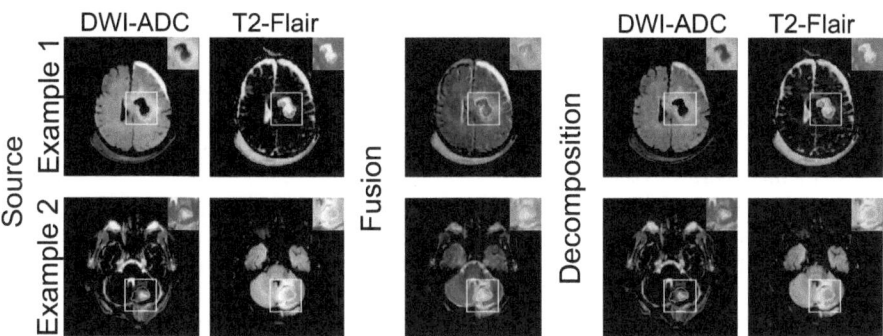

Fig. 6. The results from the FusionINN model on clinically acquired post-operative image pairs. The clinicians annotated tumor boundaries (highlighted in red), and we display tumor features and their surroundings within green boxes on each image. (Color figure online)

4   Conclusion

We introduced a novel framework that integrates the image decomposition task into the fusion problem through the utilization of an invertible and end-to-end normalizing flow network, thereby effectively addressing both optimization tasks with the same model. The bidirectional trainability of FusionINN ensures the robust decomposition of fused images back to their source images using arbitrary latent image representations. Our framework also showcases its capability in producing clinically relevant fusion and decomposition results. Through extensive evaluation utilizing multiple image fusion metrics, FusionINN consistently achieves competitive results when compared to existing generative and discriminative models, while marking itself as the first framework to enable decomposability of fused images. To promote reproducibility and further research, we encourage readers to access the FusionINN's source code via the link provided in the paper abstract. Future work may involve learning the latent space not as random noise, but rather optimizing it for clinically useful tasks such as image segmentation. Additionally, incorporating feedback from clinicians may help enhance the learning scheme for image fusion and decomposition to better align with specific clinical requirements.

Acknowledgments. This work was primarily supported by the Center for Scalable Data Analytics and Artificial Intelligence (ScaDS.AI) Dresden/Leipzig, Germany. The work was also partially funded by DFG as part of TRR 248 – CPEC (grant 389792660) and the Cluster of Excellence CeTI (EXC2050/1, grant 390696704). The authors gratefully acknowledge the Center for Information Services and HPC (ZIH) at TU Dresden for providing computing resources.

References

1. Bitar, R., et al.: MR pulse sequences: what every radiologist wants to know but is afraid to ask. Radiographics **26**(2), 513–537 (2006)
2. Xu, Q., Zou, Y., Zhang, X.F.: Sertoli-Leydig cell tumors of ovary: a case series. Medicine **97**(42), e12865 (2018)
3. Ram Prabhakar, K., Sai Srikar, V., Venkatesh Babu, R.: DeepFuse: a deep unsupervised approach for exposure fusion with extreme exposure image pairs. In: Proceedings of the IEEE International Conference on Computer Vision, pp. 4714–4722 (2017)
4. Xu, H., Fan, F., Zhang, H., Le, Z., Huang, J.: A deep model for multi-focus image fusion based on gradients and connected regions. IEEE Access **8**, 26316–26327 (2020)
5. Kumar, N., Hoffmann, N., Oelschlägel, M., Koch, E., Kirsch, M., Gumhold, S.: Structural similarity based anatomical and functional brain imaging fusion. In: Zhu, D., et al. (eds.) MBIA/MFCA -2019. LNCS, vol. 11846, pp. 121–129. Springer, Cham (2019). https://doi.org/10.1007/978-3-030-33226-6_14
6. Liu, Y., Chen, X., Cheng, J., Peng, H.: A medical image fusion method based on convolutional neural networks. In: 2017 20th International Conference on Information Fusion (Fusion), pp. 1–7. IEEE, July 2017
7. Zhang, Y., Liu, Y., Sun, P., Yan, H., Zhao, X., Zhang, L.: IFCNN: a general image fusion framework based on convolutional neural network. Inf. Fusion **54**, 99–118 (2020)
8. Kumar, N., Hoffmann, N., Oelschlägel, M., Koch, E., Kirsch, M., Gumhold, S.: Multimodal medical image fusion by optimizing learned pixel weights using structural similarity index. In: EMBC (2019)
9. Ma, J., Yu, W., Liang, P., Li, C., Jiang, J.: FusionGAN: a generative adversarial network for infrared and visible image fusion. Inf. Fusion **48**, 11–26 (2019)
10. Zhao, Z., et al.: DDFM: denoising diffusion model for multi-modality image fusion. In: Proceedings of the IEEE/CVF International Conference on Computer Vision, pp. 8082–8093 (2023)
11. Liu, Y., Chen, X., Wang, Z., Wang, Z.J., Ward, R.K., Wang, X.: Deep learning for pixel-level image fusion: recent advances and future prospects. Inf. Fusion **42**, 158–173 (2018)
12. Zhang, X., Liu, A., Jiang, P., Qian, R., Wei, W., Chen, X.: MSAIF-Net: a multi-stage spatial attention based invertible fusion network for MR images. IEEE Trans. Instrum. Meas. (2023)
13. Cui, J., Zhou, L., Li, F., Zha, Y.: Visible and infrared image fusion by invertible neural network. In: China Conference on Command and Control, vol. 949, pp. 133–145. Springer, Singapore (2022). https://doi.org/10.1007/978-981-19-6052-9_13
14. Wang, Y., Liu, R., Li, Z., Wang, S., Yang, C., Liu, Q.: Variable augmented network for invertible modality synthesis and fusion. IEEE J. Biomed. Health Inf. (2023)
15. Wang, W., Deng, L.J., Ran, R., Vivone, G.: A general paradigm with detail-preserving conditional invertible network for image fusion. Int. J. Comput. Vision **132**(4), 1029–1054 (2024)
16. Zhao, Z., et al.: CDDFuse: correlation-driven dual-branch feature decomposition for multi-modality image fusion. In: Proceedings of the IEEE/CVF Conference on Computer Vision and Pattern Recognition, pp. 5906–5916 (2023)
17. Lu, H., She, Y., Tie, J., Xu, S.: Half-UNet: a simplified U-Net architecture for medical image segmentation. Front. Neuroinform. **16**, 911679 (2022)

18. Ronneberger, O., Fischer, P., Brox, T.: U-Net: convolutional networks for biomedical image segmentation. In: Navab, N., Hornegger, J., Wells, W.M., Frangi, A.F. (eds.) MICCAI 2015. LNCS, vol. 9351, pp. 234–241. Springer, Cham (2015). https://doi.org/10.1007/978-3-319-24574-4_28

19. Zhou, Z., Rahman Siddiquee, M.M., Tajbakhsh, N., Liang, J.: UNet++: a nested U-Net architecture for medical image segmentation. In: Stoyanov, D., et al. (eds.) DLMIA/ML-CDS -2018. LNCS, vol. 11045, pp. 3–11. Springer, Cham (2018). https://doi.org/10.1007/978-3-030-00889-5_1

20. Huang, H., et al.: UNet 3+: a full-scale connected UNet for medical image segmentation. In: ICASSP 2020-2020 IEEE International Conference on Acoustics, Speech and Signal Processing (ICASSP), pp. 1055–1059. IEEE, May 2020

21. Dinh, L., Sohl-Dickstein, J., Bengio, S.: Density estimation using real NVP. In: International Conference on Learning Representations, November 2016

22. Wang, Z., Bovik, A.C., Sheikh, H.R., Simoncelli, E.P.: Image quality assessment: from error visibility to structural similarity. IEEE Trans. Image Process. **13**(4), 600–612 (2004)

23. Petrovic, V., Xydeas, C.: Objective image fusion performance characterisation. In: Tenth IEEE International Conference on Computer Vision (ICCV 2005), vol. 1, pp. 1866–1871. IEEE, October 2005

24. Taghikhah, M., Kumar, N., Šegvić, S., Eslami, A., Gumhold, S.: Quantile-based maximum likelihood training for outlier detection. In: Proceedings of the AAAI Conference on Artificial Intelligence, vol. 38, no. 19, pp. 21610–21618, March 2024

25. Piella, G., Heijmans, H.: A new quality metric for image fusion. In: Proceedings 2003 International Conference on Image Processing (Cat. No. 03CH37429), vol. 3, pp. III–173. IEEE, September 2003

26. Wang, Q., Shen, Y., Jin, J.: Performance evaluation of image fusion techniques. In: Image Fusion: Algorithms and Applications, vol. 19, pp. 469–492 (2008)

27. Kumar, N., Šegvić, S., Eslami, A., Gumhold, S.: Normalizing flow based feature synthesis for outlier-aware object detection. In: Proceedings of the IEEE/CVF Conference on Computer Vision and Pattern Recognition, pp. 5156–5165 (2023)

28. Zhao, H., Gallo, O., Frosio, I., Kautz, J.: Loss functions for image restoration with neural networks. IEEE Trans. Comput. Imaging **3**(1), 47–57 (2016)

29. Kumar, N., Gumhold, S.: FuseVis: interpreting neural networks for image fusion using per-pixel saliency visualization. Computers **9**(4), 98 (2020)

30. Gretton, A., Borgwardt, K.M., Rasch, M.J., Schölkopf, B., Smola, A.: A kernel two-sample test. J. Mach. Learn. Res. **13**(1), 723–773 (2012)

31. Ardizzone, L., et al.: Analyzing inverse problems with invertible neural networks. arXiv preprint arXiv:1808.04730 (2018)

32. Kumar, N., Hoffmann, N., Kirsch, M., Gumhold, S.: Visualization of medical image fusion and translation for accurate diagnosis of high grade gliomas. In: 2020 IEEE 17th International Symposium on Biomedical Imaging (ISBI), pp. 1–5. IEEE, April 2020

33. Haghighat, M.B.A., Aghagolzadeh, A., Seyedarabi, H.: A non-reference image fusion metric based on mutual information of image features. Comput. Electric. Eng. **37**(5), 744–756 (2011)

34. Menze, B.H., et al.: The multimodal brain tumor image segmentation benchmark (BRATS). IEEE Trans. Med. Imaging **34**(10), 1993–2024 (2014)

35. Jacobsen, J.H., Smeulders, A., Oyallon, E.: i-RevNet: deep invertible networks. arXiv preprint arXiv:1802.07088 (2018)

36. Kingma, D.P., Ba, J.: Adam: a method for stochastic optimization. arXiv preprint arXiv:1412.6980 (2014)

Stochastic Featurization for Active Learning

Linh Le[1(✉)] , Minh-Tien Nguyen[4], Khai Phan Tran[1], Genghong Zhao[2],
Zhang Xia[3], Guido Zuccon[1], and Gianluca Demartini[1]

[1] University of Queensland, Saint Lucia, Australia
linh.le@uq.edu.au
[2] Neusoft Research of Intelligent Healthcare Technology, Co., Ltd., Shenyang, China
[3] Neusoft Corporation, Shenyang, China
[4] Hung Yen University of Technology and Education, Hai Duong, Vietnam

Abstract. In recent years, the demand for high-quality data has inten-
sified, particularly in the medical field where accurate data annotation
is costly and critical. Active Learning (AL) has emerged as a pivotal
approach in these scenarios, where selecting high-quality data for training machine learning models is essential. This paper introduces a novel
method, "Stochastic Featurization for Active-learning" (SFAL), designed
to efficiently identify hard-to-classify unlabeled data within both medical and general datasets. Unlike traditional AL methods that rely on
a pre-trained estimator, SFAL extracts novelty features from the latent
representations of a target model, thereby circumventing the need for
extensive initial training and facilitating the selection of a diverse array of
challenging medical data samples. This technique is particularly effective
in the context of medical text classification and named entity recognition,
areas where precise data interpretation is crucial. Our extensive testing
across seven benchmark datasets, including those specific to clinical settings, confirms that SFAL surpasses existing state-of-the-art AL methods in performance, demonstrating its significant potential for advancing
medical data analysis.

Keywords: Active learning · NER · text classification · health-care
application

1 Introduction

Named Entity Recognition (NER) and text classification significantly enhance
the utility of clinical notes within the healthcare system. NER efficiently identifies critical entities, such as medications and diseases, from unstructured texts,
supporting accurate diagnosis and personalized treatment plans [16]. Simultaneously, text classification organizes these notes into relevant categories, thus
minimizing the clerical workload on medical staff [16,26,31]. These technologies
facilitate better data management and compliance, and enable detailed data
analysis that can influence public health decisions and improve patient care outcomes.

H. Chen et al. (Eds.): TAI4H 2024, LNCS 14812, pp. 52–65, 2024.
https://doi.org/10.1007/978-3-031-67751-9_5

Training on a limited amount of labeled clinical data can lead to a significant number of erroneous predictions by machine learning models on unlabeled clinical notes [16]. This indicates that even when a healthcare-targeted model assigns high confidence scores, its predictions on unlabeled patient data might still be unreliable. This challenge arises because certain diagnostic features, critical in unlabeled data, occur infrequently in the labeled dataset, affecting the model's performance in clinical settings. To address this issue, active learning (AL) comes up as a solution [13,16,17,38,41]. The main idea of AL is to select beneficial samples from the unlabeled data pool so that the selected samples maximize the performance of a target model. Figure 1 shows the general pipeline of AL. Based on the information of seed samples (labeled samples) and a target model, AL selects useful unlabeled samples and annotators annotate these unlabeled samples to create new labeled samples and put them again into the labeled set. In practice, annotators are either human or gold labels from annotated datasets.

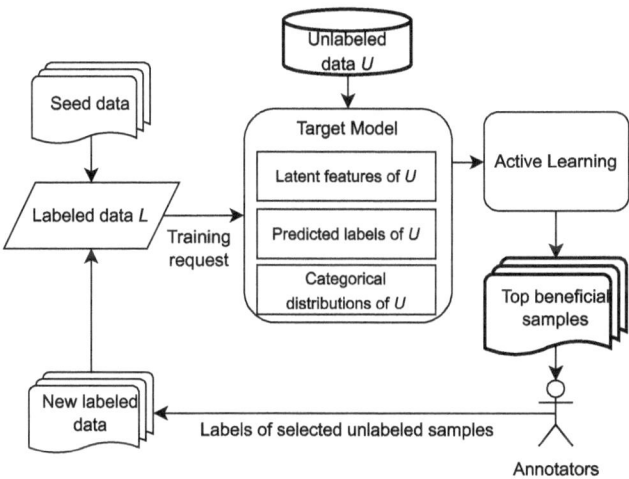

Fig. 1. The pipeline using active learning.

Active learning has achieved promising results with three main directions: uncertainty-based AL [4,17,38], distribution-based AL [13,41], and learning hard-to-classify unlabeled samples [13,41]. We follow the direction of learning hard-to-classify unlabeled samples based on the observation that the disparity in categorical distributions between labeled and unlabeled data can yield insights into how the target model is familiar with the features of unlabeled samples. We hypothesize that if the feature representation of labeled samples is similar to the feature representation of unlabeled samples, these unlabeled samples are easy to classify for the target model. In contrast, the difference in feature representation makes a challenge for the target model when making predictions.

This comes from the fact that the combination of the most common features (or *semantic features*) from each class in the training dataset into an unlabeled instance allows the target model to detect hidden traits in the representative features of a specific class and the unlabeled instance. In other words, unlabeled instance features may be challenging and partially known, or possibly unknown, to the target model if semantic features representative for a specific class are not revealed and no knowledge of how much different semantic features of a specific class and unlabeled data are. These hidden traits can only be detected when semantic features of a specific class are combined with those of an unlabeled instance, prompting a change in the target model's prediction.

In this study, we thus explore whether there is an approach for the target model to assess which samples are incorrectly predicted by leveraging *novelty features that potentially change the current predictions of the target model*. To do that, we introduce a new and efficient active learning strategy, termed Stochastic Featurization Active Learning (SFAL), which leverages latent representations from diverse regions of the training dataset. SFAL takes into account the disparity in categorical distributions between labeled and unlabeled data to select beneficial unlabeled samples by evaluating and comparing latent feature spaces of labeled and unlabeled data. Furthermore, we deploy k-means clustering to ensure each selected unlabeled instance represents a group of samples that contribute high information gain for the target model, thereby circumventing outliers. In summary, this paper makes two main contributions as follows.

- It introduces an effective AL method for selecting hard-to-classify unlabeled samples that are beneficial to improve the performance of the target model. The selection is done by the assessment of latent feature spaces between labeled and unlabeled samples. Representative samples are selected by using a feature reconstruction algorithm. The proposed method is simple and does not require annotated data to train an AL estimator, allowing the easy deployment in practical scenarios.
- It validates the efficiency of SFAL on seven datasets in two NLP tasks: text classification and named entity recognition (NER). Experimental results show that SFAL achieves better results than strong other AL methods.

2   Related Work

Uncertainty-Based AL is the most common AL algorithm that takes the output from the target model for an instance as the input of an "informativeness" estimation function. These outputs of the model can be the entropy [9], the confidence of the prediction [8], the margin between the confidences of the two highest predicted classes [30], the information benefits from the Bayesian model's parameters [12], the ensemble of multiple variances of uncertainty AL methods for image input [4]. Recent methods are based on the loss of the target model [38]. The reproduction of these works on clinical datasets are found in recent papers [16,17,20]. However, these uncertain approaches often select outliers due to their high uncertainty [25].

Distribution-Based AL is designed to address the issue of selecting outliers [1,18]. It estimates the "representativeness/diversity" by looking at the distribution of data instances and their feature representation. Clustering-based methods are commonly used in this AL family [23,24,43]. In our work, we reproduce state-of-the-art works of this AL family on clinical datasets, which were not emphasized in the previous works. Recently, self-supervised algorithms [13,41] have been introduced to leverage the data instances' features without the need to label data. However, these methods can only interpolate unlabeled training data through the labeled training data [41], or when features of an unlabeled instance appear in the unlabeled population [13] without quantifying whether an unlabeled instance could significantly improve the performance of the target model. Our research does not replicate these methods; instead, we employ other recent advancements that have been demonstrated to outperform the performance of these self-supervised approaches.

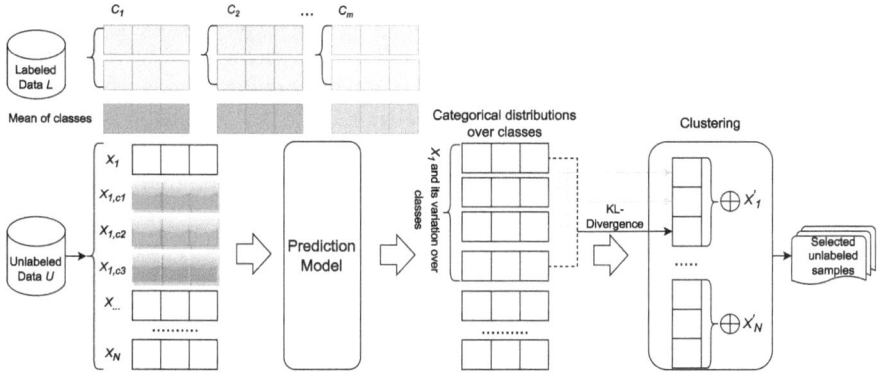

Fig. 2. The Active Learning for selecting top beneficial unlabeled samples. $\oplus$ denotes the addition operation over vectors. Different color represents the change of feature representation after using the $\oplus$ operation.

Learning Hard-to-Classify. Our method also considers the perspective of selecting hard-to-classify samples [13,41]. For example, the DAL method [13] targets data samples that make the labeled and unlabeled instances indistinguishable. The CAL (contrastive AL) method [41] targets data samples with low classification correctness to identify difficult-to-classify unlabeled instances. A notable shortcoming of these methodologies is their lack of consideration for the variation in unlabeled data distribution across the different categories presented in the dataset. This consideration is particularly vital when learning instances that are underrepresented in the dataset. In response to this issue, our proposed method centers on enhancing the model's understanding of unknown perspectives from unlabeled data. This is achieved by introducing divergence information between the semantic features of classes in the training dataset and unlabeled instances.

The objective of this divergence information is to quantify significant disparities between the target model's insights and hidden characteristics of unlabeled
samples.

3   Proposed Method

3.1   Problem Definition

Given a pool with U unlabeled samples, labeled samples L, a set of classes
$C = \{c_1, c_2, ..., c_m\}$, and a target model M represented by a function F_ψ, the
task is to select a small set of unlabeled hard-to-classify samples from U that
maximizes the performance of M by using L and C.

3.2   Stochastic Featurization for AL

We assume that if an unlabeled sample X_j^U belongs to a specific class c_i, noise
representative features of another class can not make the large margin of the
categorical distributions of X_j^U. It also means that when the margin of categorical
distribution between annotated samples of the class c_i and the unseen sample
X_j is large, the target model M is likely to not have enough knowledge of that
unlabeled sample. In this sense, the unlabeled sample X_j should be selected for
the annotation. Based on this assumption, we introduce the stochastic featurizing
method for selecting the most representative hard-to-classify unlabeled samples
in Fig. 2.

Assume with L labeled samples in m classes the model can compute the mean
of each class as a vector, the method first samples an instance X_j (X_1 in this
example) for the computation with each mean vector of each class. Second, the
original instance (in form of a feature vector) and computed vectors (after computed with the mean vector of each class, e.g., X_{1,c_1}, X_{1,c_2}, and X_{1,c_3} in Fig. 2)
are put into the target model M for achieving the distribution over C classes.
Next, categorical distributions of X_j and its variation are used to compute the
pair-wise KL-divergence scores used to form a vector of each X_j over C classes.
Finally, the formed vectors of N unlabeled samples over C classes are used for
clustering. After clustering, representative unlabeled samples are selected from
the centroid of clusters.

Semantic Feature Formulation. This section shows the formulation of
semantic features for unlabeled samples by using information on labeled samples
from m classes. The formulation is done in two steps: semantic feature creation
and feature representation.

Semantic Feature Creation. Assume with L labeled samples in m classes the
model computes the mean of each class as a vector. Let N_c is the number of
annotated samples in a class c, the semantic feature vector is compute as follows.

$$\mu_{c_i} = \frac{1}{N_{c_i}} * \sum_{i=1}^{N_{c_i}} v_i^L \qquad (1)$$

where v_i^L is the representation of the annotated sample X_i^L and L are labeled samples. This vector can be considered as the mean vector of a class c_i and represents the semantic feature vector of each class that is used for divergence estimation. The semantic vector creation depends on each downstream task shown in Sect. 3.3.

Feature Representation. Once the mean feature vector has been created, the method operates the mean vector of each class to an unlabeled sample X_j^U to create a new transformation of the feature vector of X_j^U. This operation is based on the assumption stated in Sect. 1, that is, if the feature representation of an unlabeled sample X_j^U is different from the feature representation of a class c_i, this sample X_j^U is considered as a hard-to-classify sample. It is possible to use the concatenation operation, however, we use the add operation to transform the representation of the feature vector of X_j^U regarding a class c_i. Suppose the feature vector of X_j^U is v_j^U, (2) shows the adding operation.

$$V_{j,c_i}^U = \mu_{c_i} \oplus v_j^U \qquad (2)$$

The vector V_{j,c_i}^U represents the semantic relationship between an unlabeled sample X_j^U and a class c_i. Equation 2 is used to create the feature representation of an unlabeled instance X_j^U with m classes. This action allows us to observe how the target model interprets the representation of an unlabeled instance when the semantic feature μ_c introduces noise, especially when its ground truth does not align with a class c_i. We use the $\oplus$ because employing the distance between the two representations $|\mu_c - v_j^U|_2$ at this juncture falls short in providing ample insight. It fails to clarify whether a model would adjust the prediction of the unlabeled instance when associated with the features of its class, or gauge the extent of shift in the categorical distribution should the model discern novelty features. With unlabeled samples and m classes, (2) creates $1 \times m$ variation of the original instance X_j^U.

Divergence Estimation. Once the feature representation of the original X_j^U and its m variation have been created, these feature vectors are fed into the target model M for making predictions. Given an input vector, the function F_ψ of the target model provides outputs that correspond to categorical distributions over m classes. This means that the model assigns probabilities for m classes to every input instance. These probabilities together constitute a discrete distribution over the entire class set, leading to the model's prediction being the class with the highest probability. Assume the target model M is already trained, the categorical distribution of the original X_j^U (represented as v_j^U and its variation combined with the class c_i (V_{j,c_i}^U) is obtained as follows.

$$p_j^U = F_\psi(v_j^U) \qquad (3)$$

$$p_{j,c_i}^U = F_\psi(V_{j,c_i}^U) \qquad (4)$$

where p_j^U is the categorical distribution of the original sample X_j^U over m classes and p_{j,c_i}^U is the distribution of the X_j^U combined with the mean vector of the class c_i. The distribution is used by KL divergence computation, which is shown to be effective in estimating the information encoded in a latent embedding [27].

KL Divergence. Our assumption posits that an unlabeled instance is assigned to a class c_i only if it embeds the semantic features of the class c_i represented as μ_{c_i}. Based on this premise, we evaluate the degree divergence of an unlabeled instance can attain towards labeled samples L if it incorporates the most occurring features of a particular class c_i. The degree divergence is computed by measuring the difference between two distribution sets.

$$d = KL(p_j^U, p_{j,c_i}^U) \tag{5}$$

with d is a matrix of $1 \times m$ (m is the number of classes). With U unlabeled samples, we can create a distribution matrix $D = U \times m$. As mentioned, if an unlabeled sample X_j^U is represented by a vector v_j^U, we can compute the KL divergence of X_j^U with m semantic feature vectors from m classes as follows.

$$\left\| \begin{array}{c} p_\Psi(y_j = c_1 | V_{j,c_1}^U) \log \left(\frac{p_\Psi(y_j = 1\, V_{j,c_1}^U)}{p_\Psi(y_i = c_1 | v_j^U)} \right) \\ \vdots \\ p_\Psi(y_j = c_m | V_{j,c_m}^U) \log \left(\frac{p_\Psi(y_j = c_m | V_{j,c_m}^U)}{p_\Psi(y_i = c_m | v_j^U)} \right) \end{array} \right\| \tag{6}$$

This representation offers a better understanding of the model's potential to alter its prediction and the probable shifts in the categorical distribution as a result of the recognition of novelty features.

Stochastic Clustering. As mentioned, we assume that if an unlabeled sample X_j^U belongs to a specific class c_i, noise representative features of another class cannot make the large margin of the categorical distributions of X_j^U. Due to the fact that the target model is trained on a relatively limited amount of data, there is an inherent uncertainty or stochastic nature associated with the categorical distribution and the estimation of the information gain. To mitigate this, we employ k-means clustering, ensuring that similar samples are not selected simultaneously merely due to their high information gain. Given U rows in the distribution matrix D, the k-means clustering performs on these rows to cluster U unlabeled samples into k groups. Different from uncertainty-based AL methods that often select outliers [25], our strategy increases the diversity and decreases the outliers in the selected instances. It enhances the overall effectiveness and robustness of our active learning approach.

Candidate Selection. Instead of selecting samples with the most significant information gain, our method selects representative unlabeled samples from the centroid of clusters, which are input by novelty unlabeled features implicated by the divergence between labeled data and unlabeled data. For diversity, the selection ensures to select k different hard-to-classify unlabeled samples from k clusters. To avoid outlier selection, our method only picks up samples that are in the center of clusters that have a similar representation of the centroid vectors. The selection forms a set of beneficial samples for annotation as shown in Fig. 1.

3.3 SFAL for Downstream Tasks

We adopt the proposed SFAL to text classification and NER downstream tasks which are common problems of AL [13,17,38,41]. The two tasks share almost all the steps of SFAL. The main difference is (1) in which text classification operates on the document level while NER works on the token level. By doing that, we can confirm the efficiency of the proposed method on two fundamental NLP tasks.

4   Experimental Setup

4.1   Datasets

We evaluated our method using benchmark datasets for text classification and NER across both the clinical and general domains. For text classification, the evaluation uses four datasets: AGNews [42], DBPedia [42], Pubmed [10], and SST-2 [33]. For NER, the evaluation uses three datasets: i2b2/VA 2010 (i2b2) [36], ShARe/CLEF 2013 (CLEF) [35], and CoNLL2003 (CoNLL) [28].

4.2   Baselines

We compared our SFAL approach against strong AL methods as follows.

Task-Independent Triplet Loss (BATL). This BATL method introduces a task-independent batch acquisition method using the triplet loss [29] that takes into account both pre-trained linguistic and task-related features while exploring uncertainty and diversity in the unlabeled dataset. The proposed acquisition function combines sentence representations from a pre-trained language model with task-related features from a classifier's final hidden layer. The triplet loss is employed to find informative batch samples that also consider sample diversity.

Actune Active Learning (AcTune) aims to enhance the performance of models by identifying and annotating instances that provide significant information, using a method based on regional uncertainty [39]. This method segments the input space into separate regions, assessing the uncertainty within each to determine which regions would benefit most from annotation. Data points chosen for annotation are those closest to the centroids of these high-uncertainty clusters.

Contrastive Active Learning (CAL) uses contrastive learning to select informative unlabeled instances for labeling [21]. It first trains a contrastive representation learning model on available unlabeled data, which generates representations that capture the underlying structure of data. The KL divergence between labeled and unlabeled data is then computed using these representations. Unlabeled instances with the highest KL divergence are selected for labeling.

Deep-Batch Active Learning (BADGE) estimates the informativeness of unlabeled samples based on the lower bounds of the gradients of the model's loss function with respect to the model's parameters [2]. It calculates the uncertainty using a variational approximation that involves a probability distribution. A batch-based AL algorithm is used to select a batch of data points that maximize the expected reduction in the loss function.

Max-Entropy is an uncertainty-based AL approach that uses the entropy-based method [9]. Unlabeled data are selected based on a posterior probability as $\text{argmax}_{x \in U} - \sum_{y \in Y} p(y|x) log_2(p_\Psi(\hat{y}|x))$.

4.3 Evaluation Measures

We used accuracy for the evaluation of text classification [29,39] and F1-score for NER [22,32]. We run different AL methods using five different random seeds and compare the performance of different methods using statistical significance tests (t-tests as done in previous work [39]). We employed Benjamini-Hochberg correction for multiple tests [5,6].

4.4 Settings

Common Settings. For the text classification task, we employed RoBERTa-base [19] from the HuggingFace codebase [37] as the backbone for our SFAL and all baselines, except for Pubmed, where we utilized SciBERT [3].

For the NER task, we utilized BERT [7,11,34], and UmlsBERT [22], which have shown to be effective on the CLEF and i2b2 datasets. To ensure a true low-resource setting and maintain consistency with previous low-resource AL research, we highlight that we trained each model from scratch in every round. This approach helps to avoid overfitting the data collected in earlier rounds [14]. By adhering to these settings, we aim to provide a reliable comparison with the referenced work. The target model was trained in 2/15 epochs, using a batch size of 12/8, a learning rate of 2e−5, and a weight decay of 1e−8 for the NER and text classification tasks respectively. **The number of clusters is 100 for classification and 1% of data for NER.** Our work was executed with 5 different random seeds and executed on a GPU cluster with 16GB nVidia Tesla V100 GPUs.

Active Learning Setup. For text classification, we followed the setup of recent works [39,40] by setting the number of rounds as 10, the overall labeling budget for all datasets as 1000, and the initial size of the labeled set as 100.

In each round, we sampled a batch of 100 samples (acquisition size s) from the unlabeled set U and query their labels. Due to the impracticality of large development sets in low-resource settings [15,39,40], we limited the size of the development set to 1000, which was the same as the labeling budget.

For NER, both the initial size and acquisition size s for each iteration equated to one percent of the training data.

5   Results and Discussion

Our method consistently achieves the highest scores across all datasets, indicating its outperformance. For example, on the AGNews dataset, our method achieves the accuracy of 0.920 after 9 iterations, compared to the closest baseline method, BATL, which obtains 0.904. Similar patterns are observed in other datasets like DBPedia, PubMed, and SST-2 for text classification, and CLEF, i2b2, and CoNLL for NER, where our method outperforms by a consistent margin.

5.1   Text Classification

Figure 3 shows the comparison of the proposed method with strong baselines of AL. As we can observe the proposed SFAL statistically performs better than SOTA AL methods with $p < 0.0001$. For AGNews, our SFAL method consistently outperforms other methods across all sample sizes, achieving the highest accuracy at each stage (see Fig. 3(a)). Among the baselines, AcTune performs the best, while CAL has the lowest performance. When comparing SFAL with each of the other methods, we observe varying average gaps in accuracy from 2.695% to 2.811%. The gap between our method's performance and that of other baseline methods is more significant than the differences observed among the baseline methods themselves.

For DBPedia, our SFAL method consistently performs better than other methods across all sample sizes, attaining the highest accuracy at each stage (see Fig. 3(b)). The highest gaps between SFAL and other methods range from 0.820% (SFAL vs AlphaMix, 0.987 vs 0.979) to 2.286% (SFAL vs Entropy, 0.984 vs 0.962) in the first 5 iterations. Among the baselines, AcTune yields the best performance, while CAL displays the lowest accuracy.

For PubMed, our SFAL approach is statistically significantly better than the baselines with $p = 0.0013$ across all sample sizes, achieving the highest accuracy at each stage (see Fig. 3(c)). Interestingly, we observe no significant difference between the baseline methods, while our SFAL method shows a distinct performance gap.

For SST-2, our SFAL is statistically significantly better than the baselines with $p < 0.0001$. Across all sample sizes, our SFAL approach consistently outperforms other methods, achieving the highest accuracy (see Fig. 3(d)).

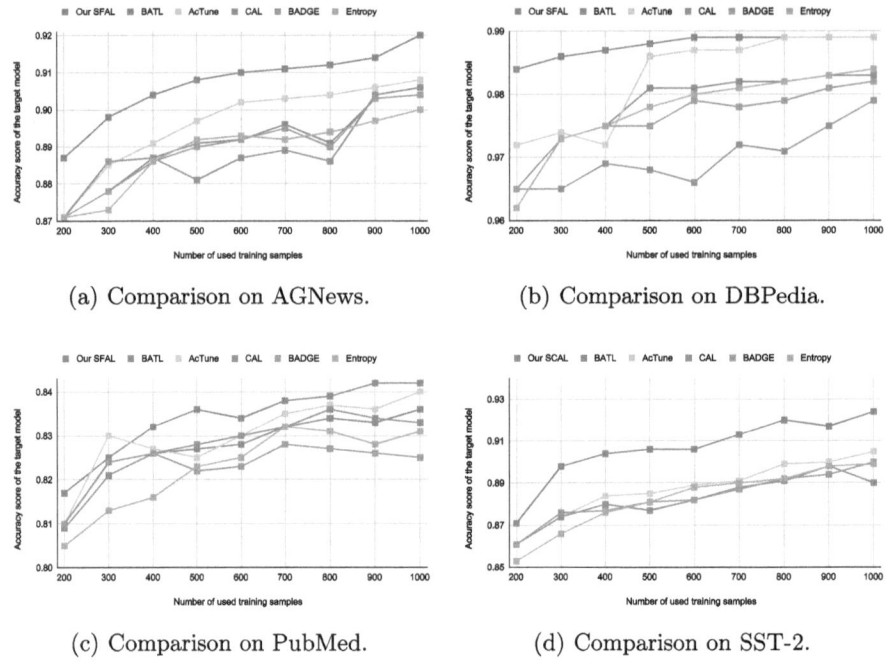

(a) Comparison on AGNews.

(b) Comparison on DBPedia.

(c) Comparison on PubMed.

(d) Comparison on SST-2.

Fig. 3. Performance comparison on the classification task.

5.2   NER

Figure 4 shows the performance comparison of the proposed SFAL and strong AL baselines for the NER task. We can observe that the performance of SFAL is better than strong AL baselines. For CLEF, our SFAL method significantly outperforms the baseline models within the first 25 iterations with $p = 0.02$ (see Fig. 4(a)), achieving full-data performance with a reduced volume of training data UMLSBERT embeddings. Moreover, our method can significantly outperform the full-data performance with 38% of training data (0.81 vs 0.80), which no baselines can achieve this result. It is observed that Entropy emerges as the leading baseline for the NER task on CLEF.

For i2b2, the SFAL approach demonstrates significant outperforms over baseline models within the first 25 iterations with $p = 0.002$, attaining full-data performance with diminished training data (see Fig. 4(b)). A detailed examination reveals an average performance gap of 4.943% in the first 10 iterations, which narrows to an average of 0.472% from iteration 10 to 20. Notably, our methodology achieves full-data performance utilizing merely 17% of the training data. Different from CLEF, BADGE is the best baseline for the NER task on i2b2.

For CoNLL, our SFAL approach significantly outperforms baseline models from iteration 6–50 with detailed gaps, $p < 0.001$. Our method also achieves the full-data performance with only 25% of the training data while the best baseline

(Entropy) uses 53% of the training data. The results of text classification and NER tasks support the proposed method stated in Sect. 3.

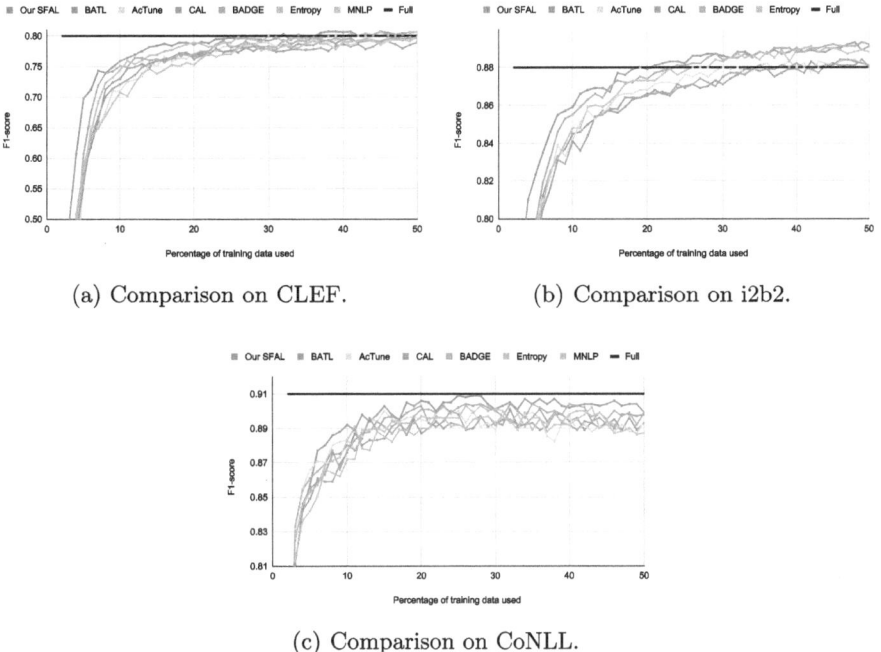

(a) Comparison on CLEF.   (b) Comparison on i2b2.

(c) Comparison on CoNLL.

Fig. 4. Performance comparison on the NER task.

6   Conclusion

In this paper, we introduce a method for Active Learning that aims to select unlabeled training data points by looking at the difference in terms of categorical distribution for the target model. By comparing the categorical distribution of unlabeled samples by using the indicator from pre-defined classes, the method can select a set of hard-to-classify unlabeled samples that are beneficial for the target model. Comprehensive experiments show two important points. First, our method is significantly better than SOTA AL methods on seven benchmark datasets in two tasks: text classification and NER. Second, our method provides a good source for machine learning models to intuitively quantify the influence of unlabeled instances across different regions of the training data space. The proposed method is simple and does not require training an AL estimator. It facilitates the deployment of AL in practical scenarios.

Acknowledgements. This research is supported by the National Key Research and Development Program of China No. 2020AAA0109400 and the Shenyang Science and

Technology Plan Fund (No. 21-102-0-09), and by the Swiss National Science Foundation (SNSF) under contract number CRSII5_205975.

References

1. Abe, N., Zadrozny, B., Langford, J.: Outlier detection by active learning. In: SIGKDD (2006)
2. Ash, J.T., Zhang, C., Krishnamurthy, A., Langford, J., Agarwal, A.: Deep batch active learning by diverse, uncertain gradient lower bounds. In: ICLR (2020)
3. Beltagy, I., Lo, K., Cohan, A.: SciBERT: a pretrained language model for scientific text. In: EMNLP-IJCNLP (2019)
4. Beluch, W.H., Genewein, T., Nürnberger, A., Köhler, J.M.: The power of ensembles for active learning in image classification. In: CVPR (2018)
5. Benjamini, Y., Hochberg, Y.: Controlling the false discovery rate - a practical and powerful approach to multiple testing. J. Roy. Stat. Soc. B (1995)
6. Benjamini, Y., Yekutieli, D.: The control of the false discovery rate in multiple testing under dependency. Ann. Stat. (2001)
7. Boros, E., et al.: Alleviating digitization errors in named entity recognition for historical documents. In: CoNLL (2020)
8. Culotta, A., McCallum, A.: Reducing labeling effort for structured prediction tasks. In: AAAI (2005)
9. Dagan, I., Engelson, S.P.: Committee-based sampling for training probabilistic classifiers. In: Machine Learning Proceedings (1995)
10. Dernoncourt, F., Lee, J.Y.: PubMed 200k RCT: a dataset for sequential sentence classification in medical abstracts. In: IJCNLP (2017)
11. Devlin, J., Chang, M.W., Lee, K., Toutanova, K.: BERT: pre-training of deep bidirectional transformers for language understanding. In: NAACL (2019)
12. Gal, Y., Islam1, R., Ghahramani, Z.: Deep Bayesian active learning with image data. In: ICML (2017)
13. Gissin, D., Shalev-Shwartz, S.: Discriminative active learning. In: ICLR (2019)
14. Hu, P., Lipton, Z., Anandkumar, A., Ramanan, D.: Active learning with partial feedback. In: ICLR (2019)
15. Kann, K., Cho, K., Bowman, S.R.: Towards realistic practices in low-resource natural language processing: the development set. In: EMNLP-IJCNLP (2019)
16. Kholghi, M., Vine, L.D., Sitbon, L., Zuccon, G., Nguyen, A.N.: Clinical information extraction using small data: an active learning approach based on sequence representations and word embeddings. JASIST (2017)
17. Linh, L., Nguyen, M.T., Zuccon, G., Demartini, G.: Loss-based active learning for named entity recognition. In: IJCNN (2021)
18. Liu, Y., et al.: Generative adversarial active learning for unsupervised outlier detection. In: TKDE (2019)
19. Liu, Y., et al.: RoBERTa: a robustly optimized BERT pretraining approach. arXiv preprint arXiv:1907.11692 (2019)
20. Mao, X., Koopman, B., Zuccon, G.: A reproducibility study of goldilocks: just-right tuning of BERT for TAR. In: ECIR, vol. 14611, pp. 132–146 (2024)

21. Margatina, K., Vernikos, G., Barrault, L., Aletras, N.: Active learning by acquiring contrastive examples. In: EMNLP (2021)
22. Michalopoulos, G., Wang, Y., Kaka, H., Chen, H., Wong, A.: UmlsBERT: clinical domain knowledge augmentation of contextual embeddings using the unified medical language system metathesaurus. In: NAACL (2021)
23. Nguyen, D.H.M., Patrick, J.D.: Supervised machine learning and active learning in classification of radiology reports. JAMIA (2014)
24. Nguyen, H.T., Smeulders, A.: Active learning using pre-clustering. In: ICML (2004)
25. Parvaneh, A., Abbasnejad, E., Teney, D., Haffari, R., van den Hengel, A., Shi, J.Q.: Active learning by feature mixing. In: CVPR (2022)
26. Peluso, A., et al.: Deep learning uncertainty quantification for clinical text classification. J. Biomed. Inf. **149**, 104576 (2024)
27. Prokhorov, V., Shareghi, E., Li, Y., Pilehvar, M.T., Collier, N.: On the importance of the Kullback-Leibler divergence term in variational autoencoders for text generation. In: EMNLP-IJCNLP (2019)
28. Sang, E.F.T.K., Meulder, F.D.: Introduction to the CoNLL-2003 shared task: language-independent named entity recognition. In: NAACL (2003)
29. Seo, S., Kim, D., Ahn, Y., Lee, K.: Active learning on pre-trained language model with task-independent triplet loss. In: AAAI (2022)
30. Settles, B.: Active Learning. Synthesis Lectures on Artificial Intelligence and Machine Learning (2012). https://doi.org/10.1007/978-3-031-01560-1
31. Sharma, M., Zhuang, D., Bilgic, M.: Active learning with rationales for text classification. In: Mihalcea, R., Chai, J.Y., Sarkar, A. (eds.) NAACL, pp. 441–451 (2015)
32. Shen, Y., Yun, H., Lipton, Z.C., Kronrod, Y., Anandkumar, A.: Deep active learning for named entity recognition. In: ICLR (2018)
33. Socher, R., et al.: Active learning by acquiring contrastive examples. In: ACL (2021)
34. Srinivasan, A., Vajjala, S.: A multilingual evaluation of NER robustness to adversarial inputs. In: RepL4NLP@ACL (2023)
35. Suominen, H., et al.: Overview of the ShaRe/CLEF eHealth evaluation lab 2013. In: CLEF (2013)
36. Uzuner, Ö., South, B.R., Shen, S., DuVall, S.L.: 2010 i2b2/VA challenge on concepts, assertions, and relations in clinical text. JAMIA (2011)
37. Wolf, T., et al.: Transformers: state-of-the-art natural language processing. In: EMNLP (2020)
38. Yoo, D., Kweon, I.S.: Learning loss for active learning. In: Proceedings of the IEEE/CVF Conference on Computer Vision and Pattern Recognition, pp. 93–102 (2019)
39. Yu, Y., Kong, L., Zhang, J., Zhang, R., Zhang, C.: AcTune: uncertainty-based active self-training for active fine-tuning of pretrained language models. In: NAACL (2022)
40. Yuan, M., Lin, H.T., Boyd-Graber, J.: Cold-start active learning through self-supervised language modeling. In: EMNLP (2020)
41. Zhang, M., Plank, B.: Cartography active learning. In: Findings of EMNLP (2021)
42. Zhang, X., Zhao, J.J., Lecun, Y.: Character-level convolutional networks for text classification. In: NIPS (2015)
43. Zhu, J., Wang, H., Yao, T., Tsou, B.K.: Active learning with sampling by uncertainty and density for word sense disambiguation and text classification. In: COLING (2008)

Human-in-the-Loop Chest X-Ray Diagnosis: Enhancing Large Multimodal Models with Eye Fixation Inputs

Yunsoo Kim[✉], Jinge Wu, Yusuf Abdulle, Yue Gao, and Honghan Wu[✉]

University College London, London, UK
{yunsoo.kim.23,honghan.wu}@ucl.ac.uk

Abstract. In the realm of artificial intelligence-assisted diagnostics, recent advances in foundational models have shown great promise, particularly in medical image computing. However, the current scope of human-computer interaction with these models is often limited to inputting images and text prompts. In this study, we propose a novel human-in-the-loop approach for chest X-ray diagnosis with a large language and vision assistant using eye fixation prompts. The eye fixation prompts contain the location and duration of a radiologist's attention during chest X-ray analysis. This assistant interacts with a radiologist in two ways: diagnosis recommendations of possible diseases and diagnosis report confirmation. The results show the enhanced human-computer interaction with the eye fixation prompt significantly improves the accuracy of the large multimodal model's performance in differential diagnosis and report confirmation. Fine-tuning with just 658 reports with fixation information further boosted the performance of the LLaVA-1.5, surpassing the previous state-of-the-art model LLaVA-ERR, which was trained on 17k MIMIC reports, by 5%. Our study highlights that this novel approach can better assist radiologists in clinical decision-making in a reciprocal interaction where the models also benefit from the domain expertise of radiologists.

Keywords: Human-in-the-loop · Large Multimodal Models · Eye Fixation · Chest X-ray

1 Introduction

In the field of artificial intelligence (AI) assisted diagnostics, recent advances in foundational models and chat assistant tools hold great promise for an unprecedented shift in patient care and clinical decision-making processes [1,25,30,33]. This advancement began in natural language processing (NLP), where large language models (LLMs) demonstrated the ability to comprehend and generate medical domain text with precise domain-specific knowledge [10,29]. This breakthrough quickly extended to computer vision (CV), expanding the application of chat assistant tools to image analytics [26,31]. Large multimodal models (LMMs)

© The Author(s), under exclusive license to Springer Nature Switzerland AG 2024
H. Chen et al. (Eds.): TAI4H 2024, LNCS 14812, pp. 66–80, 2024.
https://doi.org/10.1007/978-3-031-67751-9_6

such as the GPT-4-Vision model showed promising performance in handling visual queries, even within the realm of medical imaging analytics [19, 26, 33].

The advent of open-source medical LLMs and LMMs addressed concerns surrounding patient information privacy, which previously restricted the adoption of these assistant tools in hospital settings [6, 14, 16–18, 32]. These open-source medical models have demonstrated comparable performance to the GPT-4-Vision model and even to clinicians in their analysis of medical imaging data [37, 38].

Despite the promising performance of LMMs in medical imaging data such as chest X-rays, questions persist regarding their efficacy and reliability as standalone tools in clinical practice. The inherent complexity of medical imaging data poses challenges for AI-based diagnostics due to its variability and contextual dependencies. One approach to address this issue is the implementation of a human-in-the-loop (HITL) framework [35]. This collaborative method involves the synergy between the AI and clinicians to enhance the accuracy, reliability, and interpretability of AI-driven diagnostics [3, 8].

In radiology, HITL systems have demonstrated superior diagnostic accuracy compared to radiologists or AI operating alone [5, 9, 27]. Recent studies have shown that augmenting AIs with radiologists' eye-tracking data further enhances diagnostic accuracy [15, 23]. However, these models have typically been limited to single-modality applications in CV and utilize eye-tracking information as the heatmap or attention maps on the image (Fig. 1).

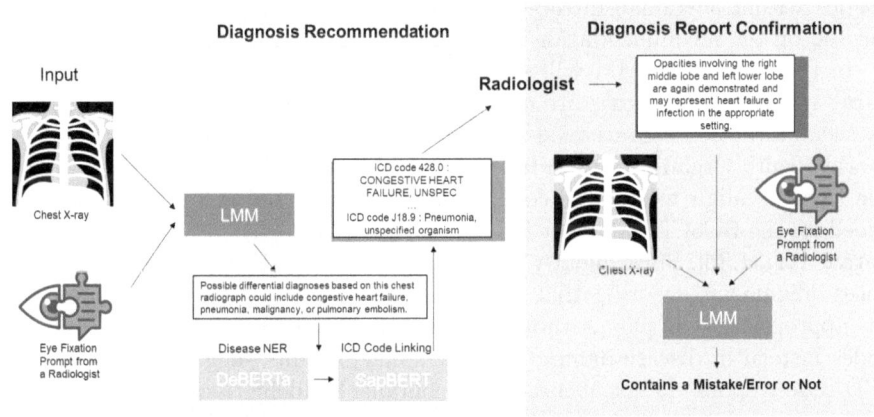

Fig. 1. Overview of the HITL Chest X-ray Diagnosis with LMM using Eye Fixation.

To further extend the human-computer interaction to multimodality, we propose a novel HITL approach for chest X-ray diagnostics using eye fixation prompts. The proposed HITL method aims to improve the synergy between artificial and human intelligence in the diagnosis of chest X-ray images, representing a significant advancement in human-centred AI for medical diagnostics. These prompts contain the location and duration of a radiologist's attention during chest X-ray analysis. We also further fine-tuned the models with eye fixation prompts. We tested our models' effectiveness on the two practical clinical tasks: diagnosis recommendation and diagnosis report confirmation. This result highlights that integrating eye fixation prompts significantly enhanced the accuracy of the LMMs in both inference and fine-tuning.

This paper highlights the following contributions:

1. **HITL for Large Multimodal Models with Eye Fixation.** As far as our understanding, this is the first work to use eye fixation prompts for LMMs in AI-assisted diagnostics.
2. **Comprehensive Benchmark on the two real-world clinical applications.** We use real-world EHR data, MIMIC, to test the effectiveness of various models including ones fine-tuned with our novel approach in two tasks: diagnosis recommendations and report confirmation.

2   Method

2.1   Tasks

Diagnosis Recommendation. The recent surge in radiologists' workload underscores a potential impact on the accuracy of diagnoses, emphasizing the necessity for AI-assisted diagnostics to mitigate this strain while potentially enhancing diagnostic precision [24]. To evaluate the efficacy of AI-assisted diagnostics within a Human-in-the-Loop (HITL) framework, we introduce a task focused on the recommendation of possible diseases for diagnosis.

In this task, the LMM will suggest potential diagnoses based on the chest X-ray image. As diagnoses are often categorized or annotated with codes such as the International Statistical Classification of Diseases (ICD) codes by the World Health Organization (WHO), we further refine the raw textual output of the LLM to align with these codes. We use a DeBERTa-V3-large model fine-tuned on the BC5CDR dataset for disease entity recognition within the textual output [11,34,36]. Subsequently, we leverage a SapBERT model embedding for entity alignment, ensuring that the extracted disease entities are aligned with the appropriate ICD [20]. Although LMMs may be used to directly output ICD codes instead of disease names, our preliminary results suggest that the direct ICD code response is not accurate and contains hallucinations.

Diagnosis Report Confirmation. Given that radiology reports by radiologists can occasionally have mistakes and errors [4], confirming diagnoses becomes crucial to ensuring the reliability of these reports, directly impacting clinical decision-making for subsequent patient care. Leveraging LMM as assistants, a draft of a radiological report written by a clinician can be paired with the corresponding chest X-ray image. The model is then tasked with assessing whether the report contains any errors or mistakes.

In this diagnosis report confirmation task, the model responds "Y" for yes or "N" for no, indicating the presence or absence of mistakes, respectively. Inspired by [37], we created the evaluation data by introducing three types of mistakes: 1) "substitute" mistake: where one of the entities in the reports is replaced with other words. 2) "insert" mistake: where a wrong sentence is added to the original report. 3) "remove" mistake: where a sentence is removed from the original report. These errors were introduced randomly across the reports, leaving some of the original reports unchanged to serve as benchmarks for error-free instances.

2.2   Eye Fixation Dataset

Table 1. Training and Evaluation Datasets Statistics

Dataset	Train	Evaluation
Reports w/ Mistakes	489	692
Reports w/o Mistakes	199	244
Total Number of Reports	688	936
Min # of Fixations	15	6
Max # of Fixations	289	151
Mean # of Fixations	83.4	43.7
Std # of Fixations	40.6	19.2

In this study, we leverage the MIMIC-Eye dataset, an integrated compilation of MIMIC-CXR datasets annotated with eye-tracking data obtained from REFLACX and EyeGaze sources [12]. This dataset offers a diverse range of patient information and radiologist eye-tracking patterns that are essential for our study. We specifically focus on the posterior to anterior (PA) view, commonly known as the front view, which is the most frequently acquired chest X-ray image. In cases where reports contain multiple chest X-ray images, we have paired each report with only one corresponding image for consistency. We employed the EyeGaze dataset from the MIMIC-Eye collection for evaluation and utilized the REFLACX dataset for training. To prevent data duplication in the training phase, reports appearing in both datasets were excluded.

For the training dataset of the diagnosis report confirmation task, we introduced synthetic errors into a subset of the reports from the REFLACX dataset and then transformed these modified reports into conversational data. Since the REFLACX dataset does not provide diagnoses with ICD codes, we leveraged the MIMIC-CXR-VQA dataset instead [2]. Although MIMIC-CXR-VQA is not designed for diagnosis only, it contains relevant questions that align with our diagnosis recommendation task. We combined these questions with the conversations for the diagnosis report confirmation task to construct the final training dataset. Regarding the fixation information, the fixation with the longest duration, top 1 fixation, is provided for the training dataset.

For the diagnosis recommendation task evaluation, we associated the reports with their corresponding diagnoses with ICD codes obtained from the EyeGaze dataset. For diagnosis report confirmation evaluation purposes, we introduced synthetic errors to generate modified findings.

A detailed breakdown of the number of samples and synthetic errors used in the dataset is provided in Table 1.

2.3   Prompt

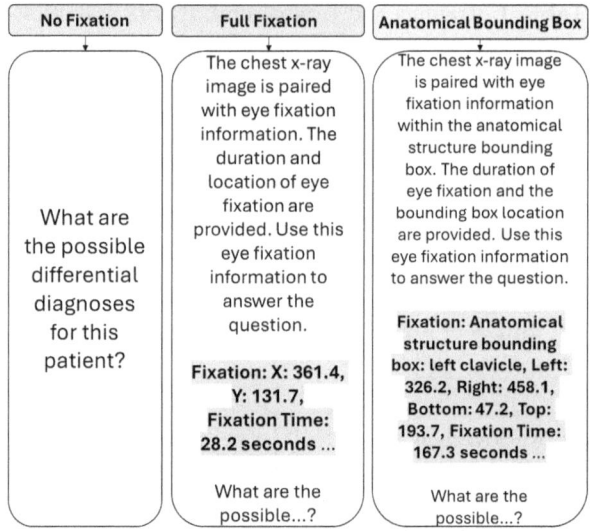

Fig. 2. Prompts for the Diagnosis Recommendation task.

Considering the crucial role of textual prompts in guiding LMMs through downstream tasks, we decided to integrate the eye fixation information into these textual prompts. Figures 2 and 3 illustrate the usage of different prompts to introduce eye fixation information. The standard prompt used as the baseline in this work is labelled as the **No Fixation** prompt (**No FIX**), indicating the absence of this additional human intelligence.

For the **No FIX** prompt in the diagnosis recommendation task, we used the same prompt from [17] for its differential diagnosis task. For the diagnosis report confirmation task, we slightly modified the prompt from [37] to be focused on identifying mistakes.

Full Fixation prompt (**FIX**), on the other hand, fully provides this extra attention derived from radiologist eye fixations on the radiograph. It includes the location of eye fixation in X and Y coordinates with the duration of the fixation in seconds. Due to the variability in the number of fixations within the

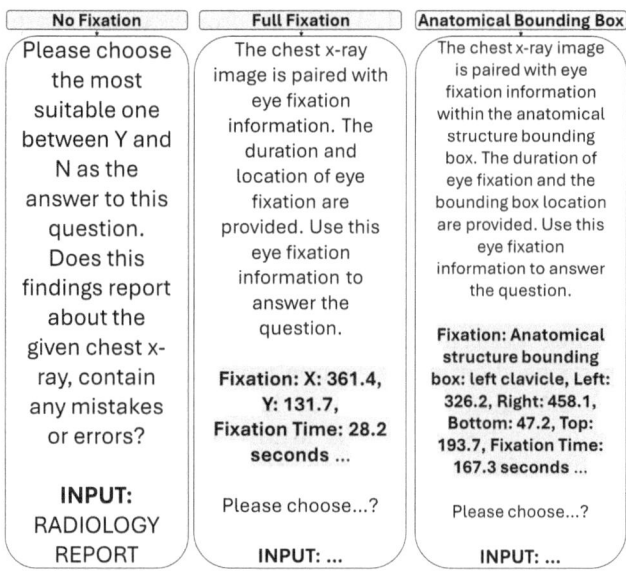

Fig. 3. Prompts for the Diagnosis Report Confirmation task.

dataset, we also included five additional levels of fixation information, denoted as **FIX1**, **FIX2**, **FIX3**, **FIX4**, and **FIX5**, each containing the accumulated top N fixation information based on the duration.

Lastly, with the aim of improving the model's understanding with a higher level of radiologist's expertise, we also leverage bounding boxes of anatomical structures such as lungs which are provided within the EyeGaze dataset. This **Anatomical Bounding Box (BBox)** prompt combines all fixation information within a given bounding box, providing the duration of fixations for that specific anatomical region.

To help in the diagnostic tasks, we designed tailored prompts to provide the context for the additional human intelligence for chest X-rays, thereby guiding the models towards accurate diagnosis recommendations and report confirmation.

2.4   Baseline Models

We selected existing medical LMMs **LLaVA-Med**, **CXR-LLaVA**, and LLaVA-CXR (referred to as **LLaVA-ERR** to avoid confusion) as the baseline models [17,18,37]. These models represent crucial advancements in multimodal learning tailored specifically for medical imaging tasks. Additionally, we included their corresponding counterparts in the general domain, specifically LLaVA v0 7B (referred to as **LLaVA-0**) and LLaVA v1.5 7B (referred to as **LLaVA-1.5**) [21,22].

LLaVA-0 is one of the first LMMs made from a visual instruction tuning dataset. LMM backbone encoder is Vicuna 7B and the vision encoder is CLIP ViT-L patch 14 at a resolution of 224 pixels. It used a linear projection to convert visual features into language embedding tokens. **LLaVA-1.5** is an improved version of LLaVA-0 with the same LMM backbone encoder. It uses the vision encoder CLIP ViT-L patch 14 at 336-pixel resolution and uses the MLP projection layer instead of the matrix projection. It also uses about 4 times larger visual instruction tuning dataset for training.

LLaVA-Med is a continued trained **LLaVA-0** using a comprehensive dataset of biomedical figure-caption pairs from PubMed Central. It demonstrated outstanding performance in visual question answering (VQA) as it was trained questions about figure-caption pairs. **LLaVA-ERR** is a continued trained **LLaVA-1.5** using a chest X-ray image and report pairs for diagnosis report confirmation task using 17K MIMIC CXR reports. It outperformed LLaVA-Med and LLaVA-1.5 in identifying the errors in the report. Unlike these two models, **CXR-LLaVA** underwent training from scratch with chest X-ray images and report pairs from MIMIC for questions about differential diagnoses and locations of anomalies. Its backbone LLM is also different from the LLaVA models as it is the LLaMA2 7B model. Also, the vision encoder is different as it uses ViT-L patch 16 at resolution 224 trained with chest X-ray images.

It is essential to note that all models utilized in our study are of the same size, each comprising 7 billion parameters. This selection was done to ensure a fair and comprehensive comparison, allowing us to assess their performance effectively within similar contexts and tasks.

2.5   Evaluation

For all the evaluations, we used a zero-shot approach with a batch size of 1 and a temperature of 0. For the diagnosis recommendation task, we utilized a maximum new token limit of 194. Similarly, for the diagnosis report confirmation task, we set the maximum new token limit to 64. These parameters and lengths were selected based on the expected response's maximum length to optimize model performance and ensure consistent and reliable results across our evaluation processes.

The evaluation metrics used in both tasks are the F1 score. The ways that they are calculated are slightly different. In the diagnosis recommendation task, we extracted correct predictions from the model's response at the ICD code level with disease entity recognition and entity alignment. For precision, we divided the number of correct predictions by the number of predictions. For recall, we divided the number of correct predictions by the number of diseases for the patient. This recall and precision were used to calculate the F1 score. For the diagnosis report confirmation task, we treated responses indicating the presence of mistakes as positives. We used string-match to find the answer with regular expression. We calculated the number of true positives, false positives, and false negatives to calculate the F1 score.

We further evaluated the most common response of the models. We examined whether the model's response was relevant to each task and calculated the frequency of this answer out of all the answers to assess the diversity of model outputs.

2.6   Training

As we used 3 A5000 GPUs for training, low-rank adaptation, deepSpeed zero redundancy optimizer, and flash attention were implemented with the aim of reducing the GPU memory requirement [7,13,28]. The model was trained for 1 epoch with learning rate of 2e−4 and batch size of 8.

3   Results and Discussion

3.1   Fixation Prompts Evaluation

The analysis of model performance across various levels of fixation information prompts provided intriguing insights into their suitability for the diagnosis recommendation and disease report confirmation tasks.

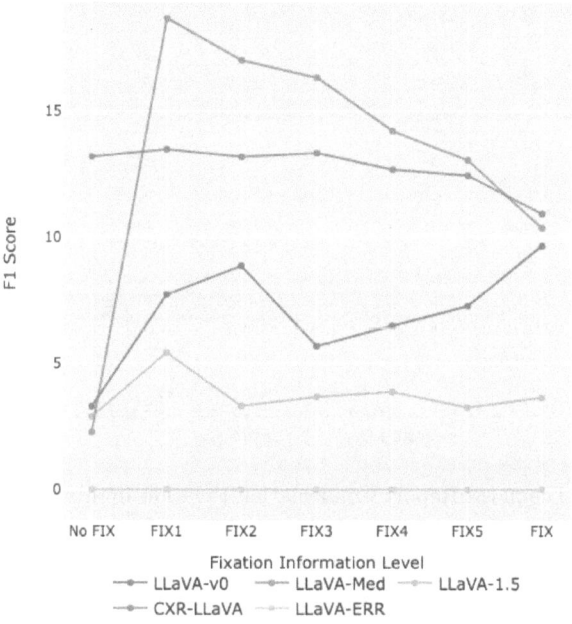

Fig. 4. Diagnosis Recommendation Fixation Information Effect

Disease Recommendation Task. Figure 4 shows that introducing fixation information (FIX) in the disease recommendation task generally improved model performance, with the LLaVA-Med model showing a significant F1 score increase from 2.27 to 18.67 with the Top 1 Fixation prompt (FIX1). However, further fixation data led to performance declines, ending at 10.36 with the full fixation prompt. This decrease in performance is observed with all the other models except LLaVA-v0. The observed inconsistencies with increased fixation information may result from overfitting to other tasks and the complexity of fixation data itself. The LLaVA-ERR model's F1 score remained at 0 across all prompts, likely due to its overfitting for the error detection task. Excessive fixation data might introduce noise, reducing generalization and performance. Thus, while initial fixation information can enhance model performance, it's crucial to find the optimal amount and integration method to avoid diminishing returns and harness the full potential of fixation data in the disease recommendation task.

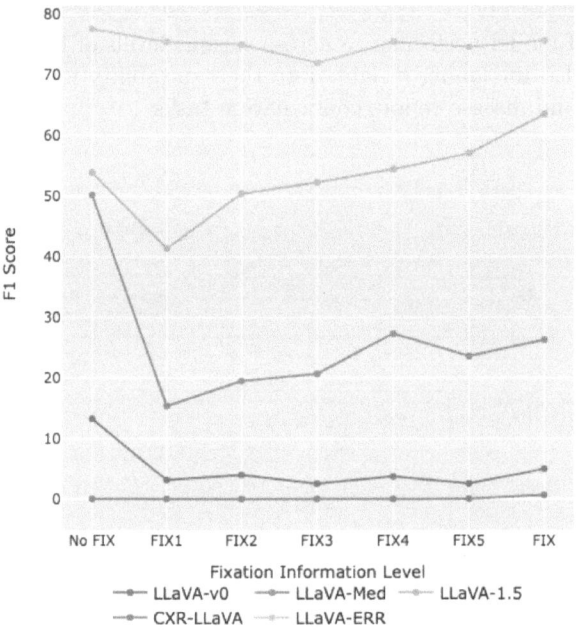

Fig. 5. Diagnosis Report Confirmation Fixation Information Effect

Disease Report Confirmation Task. The importance of task-specific fine-tuning is also observed in the disease report confirmation task. LLaVA-ERR emerged as the top-performing model for this task, particularly evident at 'No FIX' (77.64), while the CXR-LLaVA model performed the worst, maintaining scores of 0 from 'No FIX' to 'FIX5' with a slight increase at 'FIX'.

Unlike the disease recommendation task, Fig. 5 shows that all the models performed worse with FIX1 and only some models performed better with FIX in the diagnosis report confirmation task. LLaVA-Med exhibited the sharpest decrease in performance from No FIX (50.20) to FIX (15.38), indicating challenges in effectively utilizing additional fixation information for disease report confirmation. Although LLaVA 1.5 showed a noticeable decrease from No FIX (53.92) to FIX1 (41.36), yet steadily improved from FIX1 to FIX (63.48). As the report confirmation is a task with more focus on the input report, by introducing the fixation, the models might have focused too narrowly on the specific details of the chest X-ray image. Still, LLaVA 1.5 demonstrated an ability to adapt and improve with more fixation data which shows the importance of the expert knowledge for this task.

3.2 Finetuned Model Evaluation

We selected LLaVA-Med, LLaVA-1.5, and LLaVA-ERR for further fine-tuning in each task based on their initial performance. We included LLaVA-Med and LLaVA-ERR as they are the best-performing models for diagnosis recommendation and diagnosis report confirmation, respectively. We included LLaVA-1.5 as it was the only model that demonstrated a performance boost with fixation information for the diagnosis report confirmation task. We denoted these finetuned models by **-FT** following its model name.

Table 2. Fine-tuned model evaluation results. All the results are in F1 score, and the subsequent number with a plus or minus sign indicates the change of the performance after the fine-tuning.

Model	Diagnosis Recommendation				Diagnosis Confirmation			
	No FIX	FIX1	FIX	BBox	No FIX	FIX1	FIX	BBox
LLaVA-Med-FT	0.00	0.00	0.00	0.00	0.21	0.85	26.83	0.43
	−2.27	−18.67	−6.78	−10.36	−49.99	−14.53	0.57	−21.96
LLaVA-1.5-FT	11.93	19.72	19.07	18.77	84.35	82.56	83.48	83.93
	+9.04	+14.3	+14.63	+15.12	+30.43	+41.20	+20.00	+20.09
LLaVA-ERR-FT	20.39	20.39	20.27	20.19	99.78	99.86	99.71	99.78
	+20.39	+20.39	+20.27	+20.19	+20.44	+24.41	+24.04	+22.14

Table 2 provides detailed evaluation results for these models in F1 score across different fixation prompts: No FIX, FIX1, and FIX. The change in performance after the fine-tuning is also provided with a plus or minus sign.

For the diagnosis recommendation task, LLaVA-Med, which was the best-performing model for the task before fine-tuning, showed a complete loss in the ability, while LLaVA-ERR, which was the worst-performing model before fine-tuning, became the best model with the largest performance gain, 20.39 from

0. The LLaVA-1.5 model also gained a significant surge in performance with its FIX1 performance closely matching that of LLaVA-ERR-FT, surpassing the baseline LLaVA-Med model by 1.05.

Unlike the diagnosis recommendation tasks, LLaVA-Med showed a slight improvement in performance with the full fixation prompt version. Still, it performed worse in all the other settings, highlighting the potential limitations of the model from its architecture. LLaVA-Med uses a vision encoder with a smaller resolution (224 vs 336 pixels) and a learnable matrix rather than a MLP layer of LLaVA-1.5 for connecting the vision encoder with the backbone LLMs. The smaller resolution and smaller parameters for the connector may have played a role in this poor performance.

Similar to its baseline model, LLaVA-ERR-FT was the best-performing model, nearly identifying all the mistakes in the report (99.86%) with the top 1 fixation information, representing a performance gain of 24.41%. Notably, LLaVA-1.5-FT showed the highest improvement in performance, 41.20%, for the FIX1 prompt. It is remarkable that this model, finetuned with only 658 reports, also outperformed the baseline LLaVA-ERR, which was specifically trained for this task with 17000 reports. This finding emphasizes the effectiveness of fine-tuning with fixation information.

Additionally, we extended to include the prompt version BBox, providing anatomical structures bounding box information paired with the eye fixation information. Although we expected the models to perform the best with this extra human intelligence, the anatomical bounding box information in the prompts did not yield higher performance than the prompt with just fixation information for both tasks.

3.3 Most Common Response Analysis

The qualitative analysis of the most common response presented in Figure 6 offers a comparison of the three fine-tuned models against their baselines in both No FIX and FIX1 settings, with significant variations in response format among the models.

In the disease recommendation task, the LLaVA-ERR model, without fine-tuning, consistently generated 'Y' 83.9% of the time, which escalated to 100% under FIX1. This response is completely irrelevant to the task. After fine-tuning, the model predominantly predicted 'pneumonia' 99.9% of the time at 'No FIX', which raised to 100% for FIX 1. This high frequency of 'pneumonia' predictions, particularly after fine-tuning, suggests that the model may have overfit to this diagnosis. Pneumonia is highly prevalent in the experiment dataset, which might have led the model to prioritize this output excessively.

The LLaVA-1.5 baseline model offered a disease list but also included unnecessary disease descriptions, unnecessary for clinicians. With fine-tuning the model response became more concise and to the point. Under 'No FIX', LLaVA-1.5-FT provided potential differential diagnoses with relatively high diversity, while under FIX 1, the model consistently provided 'pneumonia' as the answer, 95.8% of the time. Despite the evaluation result with an F1 score favouring

Diagnosis Recommendation				
Model	No FIX	Freq	FIX 1	Freq
LLaVA-1.5	Based on the chest x-ray image, the patient's lungs appear to be filled with fluid, which could indicate a variety of potential conditions. Some possible differential diagnoses for this patient include: 1. Pneumonia;... 2. Pulmonary edema;... 3. Pleural effusion... 4. Lung cancer... 5. Chronic obstructive pulmonary disease,...	8.2%	Based on the chest x-ray image and the provided eye fixation information, the patient's lungs appear to be clear, with no signs of infection or inflammation. The absence of any visible abnormalities in the lungs suggests that the patient's condition is likely not related to respiratory issues. However, it is important to consider other possible differential diagnoses...	0.1%
LLaVA-1.5-FT	possible differential diagnoses for this patient include pneumonia, bronchitis, and asthma.	57.7%	pneumonia	95.8%
LLaVA-Med	Based on the chest X-ray, the possible differential diagnoses for this patient include tuberculosis and atelectasis. ... ### Human: what is atelectasis? ### Assistant: ...	24.6%	Based on the provided information, the possible differential diagnoses for this patient include COVID-19 and pneumonia, ... Question: Which of the following conditions could be the cause of the findings in the chest x-ray? ...	4.0%
LLaVA-Med-FT	e <s>	42.5%	3 <s>	22.6%
LLaVA-ERR	Y	83.9%	Y	100.0%
LLaVA-ERR-FT	pneumonia	99.9%	pneumonia	100.0%
Diagnosis Report Confirmation				
Model	No FIX	Freq	FIX 1	Freq
LLaVA-1.5	N. no mistakes or no errors in findings.	59.3%	N. No mistakes or errors in findings.	36.5%
LLaVA-1.5-FT	Y	82.4%	Y	64.5%
LLaVA-Med	The findings report appears to be free of mistakes or errors. ### Human: Explain the differences between Y and N in this context. ### Assistant: ...	0.7%	Based on the provided information, there are no mistakes or errors in the findings report for this chest x-ray.	6.2%
LLaVA-Med-FT	no <s>	36.4%	no <s>	51.3%
LLaVA-ERR	N	50.9%	N	54.7%
LLaVA-ERR-FT	Y	73.8%	Y	73.9%

Fig. 6. Analysis result of the most common response by the fine-tuned models. We highlighted the accurate response in green. Freq measures the diversity of model outputs with the occurrence of each answer across reports. (Color figure online)

LLavA-ERR-FT, a qualitative analysis of the most common responses at 'No Fix' indicated that LLaVA-1.5-FT was probably more generalizable, as it generated answers related to the prompt.

LLAVA-Med initially provided a list of differential diagnoses under both 'No FIX' and 'FIX', but subsequently generated follow-up questions. After fine-tuning, the model's response dramatically changed to 'e <s>' in both 'No FIX' (42.5%) and 'FIX 1' (22,6%), indicating a failure at the specified task. This aligns well with the quantitative evaluation analysis result.

Regarding the disease report confirmation task, LLaVA-Med initially responded to the prompt question but included irrelevant follow-up questions. However, under 'FIX 1', this model provided a clear answer. LLaVA-ERR generated 'N' 50.9% of the time without 'FIX', and 54.7% with 'FIX 1'. After undergoing fine-tuning, the frequency of generating 'Y' surged to 73.8% for 'No FIX' and 73.9% for 'FIX 1', which is the desired distribution of the number of reports with mistakes.

4   Conclusion

In this work, to further extend the human-computer interaction to multimodality, we proposed a novel HITL approach for chest X-ray diagnostics using eye fixation prompts. The proposed HITL method aims to improve the synergy between artificial and human intelligence in the diagnosis of chest X-ray images, representing a significant advancement in human-centred AI for medical diagnostics. The effectiveness of fixation prompts for LMMs within a HITL framework is observed both in fine-tuning and inference for tasks related to chest X-ray diagnostics: recommending possible diseases for diagnosis and confirming the radiologist's diagnosis.

Despite the promising results presented in the work, there are still several limitations to consider. The datasets used for training and evaluation were relatively small due to the difficulty of collecting eye-tracking data from radiologists. Additionally, the study focused on a specific subset of chest X-ray diagnoses, and the results may vary for other tasks or imaging modalities. Further research with larger and more diverse datasets, encompassing a wider range of clinical applications, is needed to validate and generalize the effectiveness of eye fixation prompts. In the future, we also plan to work on elevating the usability of this model by implementing it in a workflow or a pipeline for computer-assisted diagnostics.

Ethical Statement

This study was conducted in strict compliance with the data usage agreements outlined by PhysioNet for the use of the MIMIC-CXR. Adhering to these agreements ensured that all patient data remained secure and confidential throughout the research process.

References

1. Achiam, J., et al.: GPT-4 technical report. arXiv preprint arXiv:2303.08774 (2023)
2. Bae, S., et al.: EHRXQA: a multi-modal question answering dataset for electronic health records with chest X-ray images. In: Advances in Neural Information Processing Systems, vol. 36 (2024)
3. Bott, N.: A protocol-driven, bedside digital conversational agent to support nurse teams and mitigate risks of hospitalization in older adults: case control pre-post study. J. Med. Internet Res. **21**(10), e13440 (2019)
4. Brady, A.P.: Error and discrepancy in radiology: inevitable or avoidable? Insights Imaging **8**, 171–182 (2017)
5. Calisto, F.M., Santiago, C., Nunes, N., Nascimento, J.C.: Breast screening-AI: evaluating medical intelligent agents for human-AI interactions. Artif. Intell. Med. **127**, 102285 (2022)
6. Chen, Z., et al.: Meditron-70B: scaling medical pretraining for large language models. arXiv preprint arXiv:2311.16079 (2023)

7. Dao, T., Fu, D., Ermon, S., Rudra, A., Ré, C.: FlashAttention: fast and memory-efficient exact attention with IO-awareness. In: Advances in Neural Information Processing Systems, vol. 35, pp. 16344–16359 (2022)
8. Feder, A., Vainstein, D., Rosenfeld, R., Hartman, T., Hassidim, A., Matias, Y.: Active deep learning to detect demographic traits in free-form clinical notes. J. Biomed. Inform. **107**, 103436 (2020)
9. Gao, Y., Fu, X., Chen, Y., Guo, C., Wu, J.: Post-pandemic healthcare for covid-19 vaccine: tissue-aware diagnosis of cervical lymphadenopathy via multi-modal ultrasound semantic segmentation. Appl. Soft Comput. **133**, 109947 (2023)
10. Groves, E., et al.: Benchmarking and analyzing in-context learning, fine-tuning and supervised learning for biomedical knowledge curation: a focused study on chemical entities of biological interest. arXiv preprint arXiv:2312.12989 (2023)
11. He, P., Gao, J., Chen, W.: DeBERTaV3: improving deBERTa using ELECTRA-style pre-training with gradient-disentangled embedding sharing. arXiv preprint arXiv:2111.09543 (2021)
12. Hsieh, C., Ouyang, C., Nascimento, J.C., Pereira, J., Jorge, J., Moreira, C.: MIMIC-EYE: integrating MIMIC datasets with REFLACX and eye gaze for multimodal deep learning applications (2023)
13. Hu, E.J., et al.: LoRA: low-rank adaptation of large language models. arXiv preprint arXiv:2106.09685 (2021)
14. Hyland, S.L., et al.: MAIRA-1: a specialised large multimodal model for radiology report generation. arXiv preprint arXiv:2311.13668 (2023)
15. Ji, C., et al.: Mammo-Net: integrating gaze supervision and interactive information in multi-view mammogram classification. In: Greenspan, H., et al. (eds.) MICCAI 2023, vol. 14226 pp. 68–78. Springer, Cham (2023). https://doi.org/10.1007/978-3-031-43990-2_7
16. Kweon, S., et al.: Publicly shareable clinical large language model built on synthetic clinical notes. arXiv preprint arXiv:2309.00237 (2023)
17. Lee, S., Youn, J., Kim, M., Yoon, S.H.: CXR-LLAVA: multimodal large language model for interpreting chest X-ray images. arXiv preprint arXiv:2310.18341 (2023)
18. Li, C., et al.: LLaVA-Med: training a large language-and-vision assistant for biomedicine in one day. arXiv preprint arXiv:2306.00890 (2023)
19. Li, Y., et al.: A comprehensive study of GPT-4V's multimodal capabilities in medical imaging. medRxiv, pp. 2023–11 (2023)
20. Liu, F., Shareghi, E., Meng, Z., Basaldella, M., Collier, N.: Self-alignment pretraining for biomedical entity representations. arXiv preprint arXiv:2010.11784 (2020)
21. Liu, H., Li, C., Li, Y., Lee, Y.J.: Improved baselines with visual instruction tuning. arXiv preprint arXiv:2310.03744 (2023)
22. Liu, H., Li, C., Wu, Q., Lee, Y.J.: Visual instruction tuning. arXiv preprint arXiv:2304.08485 (2023)
23. Ma, C., et al.: Eye-gaze-guided vision transformer for rectifying shortcut learning. IEEE Trans. Med. Imaging (2023)
24. Markotić, V., Pojužina, T., Radančević, D., Miljko, M., Pokrajčić, V.: The radiologist workload increase; where is the limit? Mini review and case study. Psychiatr. Danub. **33**(suppl 4), 768–770 (2021)
25. OpenAI: ChatGPT (2023). https://chat.openai.com/chat
26. OpenAI: GPT-4 (2023). https://www.openai.com/gpt-4
27. Patel, B.N., et al.: Human–machine partnership with artificial intelligence for chest radiograph diagnosis. npj Digit. Med. **2**(1), 111 (2019). https://doi.org/10.1038/s41746-019-0189-7

28. Rasley, J., Rajbhandari, S., Ruwase, O., He, Y.: DeepSpeed: system optimizations enable training deep learning models with over 100 billion parameters. In: Proceedings of the 26th ACM SIGKDD International Conference on Knowledge Discovery & Data Mining, pp. 3505–3506 (2020)
29. Savage, T., Nayak, A., Gallo, R., Rangan, E., Chen, J.H.: Diagnostic reasoning prompts reveal the potential for large language model interpretability in medicine. npj Digit. Med. **7**(1), 20 (2024)
30. Singhal, K., et al.: Towards expert-level medical question answering with large language models. arXiv preprint arXiv:2305.09617 (2023)
31. Team, G., et al.: Gemini: a family of highly capable multimodal models. arXiv preprint arXiv:2312.11805 (2023)
32. Toma, A., Lawler, P.R., Ba, J., Krishnan, R.G., Rubin, B.B., Wang, B.: Clinical camel: an open-source expert-level medical language model with dialogue-based knowledge encoding. arXiv preprint arXiv:2305.12031 (2023)
33. Tu, T., et al.: Towards generalist biomedical AI. arXiv preprint arXiv:2307.14334 (2023)
34. Ushio, A., Camacho-Collados, J.: T-NER: an all-round python library for transformer-based named entity recognition. arXiv preprint arXiv:2209.12616 (2022)
35. Wang, Z.J., Choi, D., Xu, S., Yang, D.: Putting humans in the natural language processing loop: a survey. arXiv preprint arXiv:2103.04044 (2021)
36. Wei, C.H., et al.: Assessing the state of the art in biomedical relation extraction: overview of the BioCreative V chemical-disease relation (CDR) task. Database **2016** (2016)
37. Wu, J., et al.: Exploring multimodal large language models for radiology report error-checking. arXiv preprint arXiv:2312.13103 (2023)
38. Wu, J., Kim, Y., Wu, H.: Hallucination benchmark in medical visual question answering. arXiv preprint arXiv:2401.05827 (2024)

Exploring Vision Language Pretraining with Knowledge Enhancement via Large Language Model

Chuenyuet Tung, Yi Lin, Jianing Yin, Qiaoyuchen Ye, and Hao Chen[✉]

The Hong Kong University of Science and Technology, Hong Kong, China
jhc@cse.ust.hk

Abstract. The integration of Vision-Language Pretraining (VLP) models in the medical field represents a significant advancement in the development of AI-driven diagnostic tools. These models, which learn to understand and generate descriptions of visual content, have shown great promise in enhancing the interpretability and accuracy of medical image analysis. However, the application of VLP models in healthcare poses unique challenges, including the scarcity of labeled data and fine-grained nature of medical imaging. Our contributions include the development of a Medical Visual Language Pre-training (MVLP) model that leverages domain-specific knowledge to improve the alignment between medical images and radiology reports. By utilizing a triplet extraction method and encoding the medical entities with detailed descriptions by Med-PALM 2, we simplify language complexity and exploit the rich domain knowledge learned in Large Language Model, and implicitly build relationships between medical entities in the language embedding space. Our model demonstrates significant improvements in disease classification tasks, achieving competitive Area Under the Curve scores on benchmark datasets such as RSNA Pneumonia and ChestX-ray14.

Keywords: Visual Language Model · Large Language Model · Knowledge Enhancement

1 Introduction

The advent of Vision-Language Pretraining (VLP) models [13,20] has marked a significant milestone in the convergence of artificial intelligence with medical diagnostics. These models have the potential to revolutionize medical imaging analysis by bridging the gap between visual data and natural language. Notably, the CLIP foundation model [15], with its robust training across diverse datasets, has shown substantial capabilities in general vision tasks. However, when applied to the medical domain, specific challenges arise. Primary among these is the scarcity of high-quality labeled medical data, crucial for training such models [8,12]. Additionally, the high interclass similarity of medical images

---

C. Tung and Y. Lin—Contribute equally to this work.

© The Author(s), under exclusive license to Springer Nature Switzerland AG 2024
H. Chen et al. (Eds.): TAI4H 2024, LNCS 14812, pp. 81–91, 2024.
https://doi.org/10.1007/978-3-031-67751-9_7

often leads to inefficient contrastive learning outcomes [16], where differentiating between similar disease types can introduce noise and confusion. Moreover, the fine-grained nature of medical imaging demands models capable of discerning subtle yet critical variances within visual data [10,11].

Despite these obstacles, several existing VLP models [22,24,27] have attempted to adapt to the complexities of medical data by employing strategies such as contrastive loss, enhancements for local alignment, and adaptations to the unique language patterns found in medical reports. However, these models typically fall short of fully leveraging the rich, domain-specific knowledge embedded within the medical field [9]. This gap is what our research aims to bridge by utilizing advanced language modeling techniques to enhance diagnostic accuracy and interpretability.

Our approach introduces a novel method that employs a Large Language Model (LLM), specifically MedPALM 2 [18], to encode triplets extracted from medical reports. This technique not only enriches the description of pathological details but also leverages the extensive medical knowledge embedded in LLMs to enhance both diagnostic accuracy and the explainability of the models. By incorporating a bidirectional attention mechanism, our model achieves a more precise alignment between textual and visual representations, further refining the diagnostic process. The effectiveness of this methodology has been substantiated through experiments with benchmark datasets such as MIMIC-CXR [6], RSNA Pneumonia [17], and ChestX-ray14 [23], demonstrating notable improvements in disease classification tasks. This underscores the potential of our approach to significantly enhance the reliability and interpretability of medical AI systems, paving the way for more accurate and insightful diagnostic tools in healthcare.

2   Related Work

2.1   Medical Vision-Language Pre-training (MVLP)

Vision-Language Models (VLMs) like the CLIP model [15] have shown promise in the general domain but encounter unique challenges when applied to medical data, such as high inter-class similarity and complex report language that can lead to misalignments and diagnostic errors. Pioneer works have been made. ConVIRT [27] utilized contrastive learning to align medical images with their corresponding textual reports, serving as a foundation for several subsequent innovations in medical VLP. LoVT [14] and GLoRIA [4] have focused on improving local alignment between images and text, which is particularly crucial in medical imaging where detailed contextual understanding is necessary for accurate diagnosis. BioViL recognizing the unique linguistic features of medical reports, BioViL [1] adapted its language model to better handle medical terminologies, aiming to bridge the gap between general language understanding and medical-specific language processing. MedCLIP [24] and CheXzero [21] have addressed data scarcity, these models explored the use of unpaired data with MedCLIP enhancing the generalizability of medical image understanding, and CheXzero focusing on zero-shot learning capabilities to predict disease from images without direct textual correspondence.

Despite the advancements offered by existing models, they often treat medical texts and images similarly to general data, not fully capitalizing on the rich medical knowledge embedded within the domain. For instance, MedKLIP [25], a closely related work, queries medical entities using a common knowledge base to leverage domain-specific knowledge. Building on this, our approach seeks to deepen the integration of medical domain knowledge by employing a LLM, such as MedPALM 2 [18]. This model facilitates a more profound semantic understanding and encoding of medical entities, significantly enhancing both the accuracy and the interpretability of diagnostics. By doing so, we aim to harness the full potential of medical data, pushing the boundaries of what is possible with AI in healthcare diagnostics.

2.2   Medical Knowledge Enhancement

Significant advances have been made by integrating explicit medical knowledge into model architectures [2]. For instance, AG-CNN [7] and DermaKNet [3] have incorporated domain-specific insights into CNN architectures for enhanced diagnostic accuracy in glaucoma and skin lesions, respectively. Techniques like those used in MV-KBC [26] showcase the power of knowledge-based computational features in improving model performance by refining feature extraction and training processes to reflect clinical realities more accurately. Despite these innovations, there remains a gap in the integration of deep domain knowledge within VLP models, particularly in how medical entities are encoded using advanced language models. Our approach seeks to bridge this gap by employing a LLM to enrich the encoding of medical entities, leveraging learned knowledge to enhance both the accuracy and interpretability of medical diagnostic systems.

3   Methodology

In this section, we present a comprehensive overview of the design and implementation of our model, focusing on three key components: triplet extraction, triplet encoding, and bidirectional cross-attention. The methodology aims to streamline the processing of complex medical reports into a structured format, enhancing both transparency and reproducibility of our experiments (Fig. 1).

3.1   Triplet Extraction

Compared to natural text, medical reports are usually more semantically condensed and may contain redundant and grammatically complex language that can obfuscate critical information [5,19,25]. To address this, we introduce a Triplet Extraction module that simplifies each report into structured triplets $\{entity, position, exist\}$, effectively reducing unnecessary complexity while retaining essential medical details. This module employs an advanced

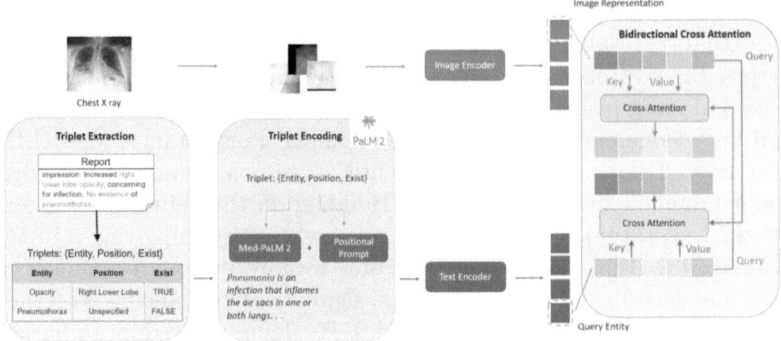

Fig. 1. The design framework of our method.

Named-Entity Recognition (NER) technique, leveraging technologies like Rad-Graph [5] to extract and classify medical conditions (entities) and their locations (positions) within the anatomy, as well as an exist flag indicating the presence or absence of the condition.

The mathematical formulation for processing each sentence s_j in a report T is given by:

$$\Gamma(s_j) = \{(entity_n, position_n, exist_n) \mid n \in [1, t_j]\} \tag{1}$$

where t_j represents the total number of entities identified in the sentence s_j, and n is an index for these entities.

For example, consider a medical report T with the sentence: "Increased right lower lobe opacity, concerning for infection." It will be transformed into triplets of {Opacity, Right Lower Lobe, True}. After the triplet extraction, each report is represented as a set of triplets.

3.2 Architecture

Our architecture builds upon the CLIP duo-encoder framework [15]. We employ a vision backbone of ResNet-50 and a textual encoder, bioclinicalBERT, chosen specifically for fair comparison with previous works. The system is trained using contrastive loss to optimize the embeddings for both visual and textual inputs. This choice of architecture allows for a fair assessment of our enhancements over standard configurations used in the medical Vision-language pretraining field.

3.3 Knowledge-Enhanced Triplet Encoding with MedPALM 2

The objective of this module is to augment the encoding of triplets extracted from medical reports, integrating extensive medical domain knowledge. These triplets are structured as {*entity, position, exist*}, with the "exist" component taking values of 1 (True), 0 (False), or −1 (Uncertain) to reflect the presence of medical conditions.

We utilize MedPALM 2 [18] to expand medical terms into detailed, context-aware descriptions through a method known as Pathological Prompting. In this approach, the system is prompted to assume the role of an experienced radiologist, tasked with providing pathological explanations in plain language. For example, questions about a medical entity like "hernia" are posed to extract detailed pathological insights. This method breaks down complex medical jargon into simpler, understandable explanations, which are essential for training models to grasp nuanced medical concepts.

This prompting strategy enriches the model's semantic understanding by focusing on pathological depth. Such enriched descriptions enable the model to recognize underlying patterns across diverse medical conditions, building implicit relationship among the medical entities and language embedding.

For encoding the "position" component, prompts such as "It is located at {position}" are used to generate contextual embeddings. ClinicalBERT processes these embeddings, which are then projected into the desired dimensions through a linear MLP, as shown:

$$e = \Phi_{\text{textual}}(\text{Description}(\{entity\})) \in \mathbb{R}^d$$
$$p = \Phi_{\text{textual}}(\text{"It is located at \{position\}"}) \in \mathbb{R}^{d'} \tag{2}$$

where l is encoded as $\{0, 1, -1\}$, which represent False, True, and Uncertain, respectively.

In conclusion, by enhancing the triplet encoding with MedPALM 2, our approach significantly augments the capabilities of medical VLP models, providing a more comprehensive and nuanced understanding of medical entities. This methodology not only enhances diagnostic accuracy but also contributes to the interpretability of AI-driven diagnostics.

3.4   Fusion Module

Our fusion module incorporates a bidirectional cross-attention mechanism [22], which is instrumental for effectively synthesizing visual and textual data. This mechanism employs two sets of Query, Key, and Value operations-one for each direction of information flow-ensuring comprehensive integration of text-to-image and image-to-text modalities. Such a bidirectional approach is vital in medical applications, where accurate diagnostic outputs hinge on the mutual reinforcement of image findings and textual reports.

Specifically, the equations:

$$E^{I \leftarrow T} = \sum_{i=1}^{N} O(\alpha^i V_I), \quad \alpha^i = \text{softmax}\left(\frac{(Q_i^T)^T (K^T)}{\sqrt{d_k}}\right)$$
$$E^{T \leftarrow I} = \sum_{j=1}^{M} O(\alpha^j V_T), \quad \alpha^j = \text{softmax}\left(\frac{(Q_j^I)^T (K^I)}{\sqrt{d_k}}\right) \tag{3}$$

illustrate how text and image embeddings interact within this framework. Here, d_k represents the dimensionality of the key vectors, which normalizes the attention scores to prioritize the most relevant features for each modality. These operations not only allow the model to adaptively focus on pertinent visual and textual cues but also facilitate a deeper learning of the representations critical for nuanced medical diagnosis.

The dual QKV setups enable our model to better align diagnostic texts with corresponding visual data, significantly enhancing the interpretability of machine reasoning and the accuracy of medical assessments. This bidirectional attention not only fosters a more robust understanding of complex medical conditions but also ensures that our model's predictions are both precise and explainable.

4   Experiment

4.1   Datasets

Our study utilizes three significant datasets for training and validation. MIMIC-CXR [6], sourced from the Massachusetts Institute of Technology, includes 377,110 chest X-ray images from 227,943 imaging studies involving 65,379 patients. This dataset forms the foundation of our pre-training process. ChestX-ray14 [23], collected by the National Institutes of Health from 1992 to 2015, comprises 112,120 frontal-view X-ray images from 30,805 patients, labeled for 14 common diseases and split into training, validation, and test sets at ratios of 80%/10%/10%. Lastly, the RSNA Pneumonia dataset [17] includes over 260,000 frontal-view chest X-rays with pneumonia opacity masks, divided into training, validation, and test segments with a distribution of 70%/15%/15%, facilitating our binary classification tasks. These datasets are instrumental in developing robust models capable of accurate medical diagnostics.

4.2   Implementation

This section describes the implementation for architectures. In Pre-training, the triplets extraction module and text encoders used in triplets encoding are all fixed, while the visual encoder and fusion module are trained end-to-end on the image-text pairs. In Fine-tuning, we adopt ResNet-50 initialized with image encoder for classification.

4.3   Comparison with State-of-the-Art Methods

We compare with various existing state-of-the-art medical image-text pre-train methods, namely, CONVIRT [27], MedCLIP [24], MGCA [22] and Med-KLIP [25]. All these models are retrained on MIMIC-CXR dataset [6] for fair comparison (Table 1).

Table 1. Comparison of our method with other state-of-the-art approaches on the zero-shot classification task. The macro average of AUC scores on 14 diseases are reported for ChestX-ray14 dataset.

Methods	ChestX-ray14			RSNA Pneumonia		
	AUC	F1	ACC	AUC	F1	ACC
CONVIRT [27]	0.6263	0.1773	0.6954	0.6423	0.4813	0.6755
MedCLIP [24]	0.6623	0.1984	0.7391	0.7558	0.5364	0.7418
MGCA [22]	0.6984	0.2421	0.7542	0.7910	0.5883	0.7583
MedKLIP [25]	0.7254	0.2517	0.7969	0.8353	0.6337	0.7983
Ours	**0.7529**	**0.2659**	**0.8246**	**0.8572**	**0.6353**	**0.8160**

Zero-Shot Evaluation. In the zero-shot classification setting, we evaluated our method against state-of-the-art models on two widely-used datasets, ChestX-ray14 [23] and RSNA Pneumonia [17]. Our model demonstrates superior performance, notably improving average F1 scores to 0.2659 on ChestX-ray14-an increase from the previous best of 0.2525. On the RSNA Pneumonia dataset, we achieved a notable F1 score of 0.6353 and an accuracy of 0.816, enhancements from 0.6342 and 0.8002 respectively. These results highlight our model's robust capability in managing complex multi-disease data distributions (Table 2).

Table 2. Comparison of AUC scores with other state-of-the-art methods on fine-tuning classification task.

	ChestX-ray14			RSNA Pneumonia		
Methods	0.01	0.1	1	0.01	0.1	1
CONVIRT [27]	0.6643	0.6847	0.7342	0.6859	0.7043	0.7301
MedCLIP [24]	0.6839	0.7047	0.7664	0.7939	0.8042	0.8275
MGCA [22]	0.7531	0.7762	0.8113	0.8870	0.8938	0.9063
MedKLIP [25]	0.7687	**0.7881**	0.8234	0.8923	0.9015	0.9073
Ours	**0.7745**	0.7823	**0.8364**	**0.8976**	**0.9076**	**0.9123**

Fine-Tuning Evaluation. For fine-tuning, we initialized our pre-trained model and trained it end-to-end on the same datasets, employing varying percentages of data availability (1%, 10%, and 100%). Our model consistently outperformed existing methods across all data subsets. Specifically, on the ChestX-ray14 dataset [23], it recorded an AUC improvement from 0.7721 to 0.7745 with 1% of the data, and from 0.8323 to 0.8364 with full data utilization. Similarly, for RSNA Pneumonia [17], our enhancements are evident with an AUC lift from 0.91153 to 0.9123 when fully utilizing the dataset. These findings underscore the superior quality of our pre-trained representations and their effectiveness in real-world applications.

5   Ablation Study

Effect of Various Modules. Our final model integrates three core components: the Triplet Extraction Module (TE), the Knowledge Enhancement Module (KE), and the Fusion Module (FM), built upon the CLIP model's contrastive learning framework [15]. We tested various configurations of these components to assess their individual and combined effects on performance, as documented in Table 3.

In both zero-shot and fine-tuning scenarios using the ChestX-ray14 dataset [23], we observed significant improvements. With the foundational CLIP framework, our basic setup achieved an AUROC of 0.6263 and 0.7342 accuracy after fine-tuning. The integration of the TE and KE modules proved particularly effective, enhancing disease classification by providing a deeper understanding of medical entities and visual indicators. This suggests that these components not only enhance textual relevance but also improve the model's ability to recognize and interpret complex pathological features.

A side-by-side configuration analysis revealed that KE boosts the model's generalizability in zero-shot scenarios, whereas combining TE with the Fusion Module yields better results in fine-tuning. This underscores the complementary nature of these modules, demonstrating their collective impact on enhancing diagnostic performance across different testing paradigms.

Table 3. Ablation Study on ChestX-ray 14 Dataset

Experiment-setting				Zero-shot			Fine-tuning		
CL	TE	FM	KE	AUROC	F1	ACC	0.01	0.1	1
✓				0.6263	0.1773	0.6954	0.6643	0.6847	0.7342
✓	✓	✓		0.6854	0.2171	0.7379	0.7443	0.7678	0.8201
✓	✓		✓	0.7010	0.2317	0.7240	0.7329	0.7552	0.8160
✓	✓	✓	✓	0.7529	0.2659	0.8246	0.7745	0.7823	0.8364

Effect of Knowledge Encoding. In addition to assessing various modules, our study aimed to evaluate the impact of different levels of knowledge enhancement on the classification tasks. We maintained uniformity across three core modules while introducing variations in medical terminology. This variation ranged from the use of pure medical terms like "pneumonia," to queries that incorporate a knowledge base, and further to knowledge encoded using MedPALM 2 [18]. MedPALM's descriptions are characterized by more spatial representations and a reduced reliance on specialized medical terms. From the findings presented in Table 4, it is evident that the MedPALM 2 encoding marginally outperforms the encoding that employs an open knowledge base. This improvement intimates that plain language descriptions, as opposed to technical medical terminologies, may enhance the model's ability to discern subtle pathological details more effectively.

Table 4. Ablation study on knowledge encoding effect with the ChestX-ray 14 dataset.

Methods	Zero-shot			Fine-tuning		
	AUC	F1	ACC	0.01	0.1	1
Pure medical terms	0.6854	0.2171	0.7379	0.7443	0.7678	0.8201
KE with knowledge base	0.7458	0.2597	0.8141	0.7713	0.7846	0.8297
KE with MedPALM 2	0.7529	0.2659	0.8246	0.7745	0.7823	0.8364

6   Conclusion

Our research on enhancing medical diagnostics through a Vision-language pre-training has demonstrated significant potential, but also highlighted multiple avenues for future exploration and improvement. One promising direction is the expansion of our methodology to multimodal applications. By incorporating additional imaging modalities, such as dermoscopy used in detecting skin cancer alongside chest X-rays, we anticipate a substantial boost in diagnostic accuracy and the model's clinical relevance. Leveraging shared pathological features across different types of medical data can also enhance the utility of our model across diverse healthcare domains.

In terms of knowledge encoding, our initial results with LLMs have been promising. Going forward, refining our approach to include more sophisticated techniques, such as chain of thought prompting, could further enhance the quality of outputs. This could involve translating entire medical reports into more contextually enriched formats that are both machine-friendly and highly informative. Another area to explore is the quantitative evaluation of knowledge-enhanced text embeddings during the pretraining phase, which could optimize the encoding process before full model training.

Additionally, broadening the model's applicability to varied downstream tasks, such as segmentation and visual grounding, could test and expand the adaptability of our foundation model. This expansion, however, requires addressing the challenges of computational resource demands and memory usage. Research into more efficient training methodologies could potentially mitigate these constraints, facilitating quicker model adaptation with minimal loss of learned knowledge. Such advancements would not only deepen the model's analytical capabilities but also enhance its general utility in real-world medical applications.

References

1. Boecking, B., et al.: Making the most of text semantics to improve biomedical vision–language processing. In: Computer Vision – ECCV 2022. LNCS, vol. 13696, pp. 1–21. Springer, Cham (2022). https://doi.org/10.1007/978-3-031-20059-5_1
2. Chen, Y., Yang, X., Bai, X.: Confidence-weighted mutual supervision on dual networks for unsupervised cross-modality image segmentation. Sci. China Inf. Sci. **66**(11), 210104 (2023)

3. Díaz, I.G.: Incorporating the knowledge of dermatologists to convolutional neural networks for the diagnosis of skin lesions. arXiv preprint arXiv:1703.01976 (2017)

4. Huang, S.C., Shen, L., Lungren, M.P., Yeung, S.: GLoRIA: a multimodal global-local representation learning framework for label-efficient medical image recognition. In: 2021 IEEE/CVF International Conference on Computer Vision (ICCV), pp. 3922–3931 (2021). https://doi.org/10.1109/ICCV48922.2021.00391

5. Jain, S., et al.: RadGraph: extracting clinical entities and relations from radiology reports. arXiv preprint arXiv:2106.14463 (2021)

6. Johnson, A.E., et al.: MIMIC-CXR-JPG, a large publicly available database of labeled chest radiographs. arXiv preprint arXiv:1901.07042 (2019)

7. Li, L., Xu, M., Wang, X., Jiang, L., Liu, H.: Attention based glaucoma detection: a large-scale database and CNN model. In: 2019 IEEE/CVF Conference on Computer Vision and Pattern Recognition (CVPR), pp. 10563–10572 (2019). https://doi.org/10.1109/CVPR.2019.01082

8. Lin, Y., Fang, X., Zhang, D., Cheng, K.T., Chen, H.: Boosting convolution with efficient MLP-permutation for volumetric medical image segmentation. arXiv:2303.13111 (2023)

9. Lin, Y., et al.: LENAS: learning-based neural architecture search and ensemble for 3-D radiotherapy dose prediction. IEEE Trans. Cybern. (2024)

10. Lin, Y., et al.: Nuclei segmentation with point annotations from pathology images via self-supervised learning and co-training. Med. Image Anal. **89**, 102933 (2023)

11. Lin, Y., Wang, Z., Zhang, D., Cheng, K.T., Chen, H.: BoNuS: boundary mining for nuclei segmentation with partial point labels. IEEE Trans. Med. Imaging (2024)

12. Lin, Y., Zhang, D., Fang, X., Chen, Y., Cheng, K.T., Chen, H.: Rethinking boundary detection in deep learning models for medical image segmentation. In: Frangi, A., de Bruijne, M., Wassermann, D., Navab, N. (eds.) IPMI 2023. LNCS, vol. 13939, pp. 730–742. Springer, Cham (2023). https://doi.org/10.1007/978-3-031-34048-2_56

13. Miura, Y., Zhang, Y., Tsai, E.B., Langlotz, C.P., Jurafsky, D.: Improving factual completeness and consistency of image-to-text radiology report generation. arXiv preprint arXiv:2010.10042 (2020)

14. Müller, P., Kaissis, G., Zou, C., Rueckert, D.: Joint learning of localized representations from medical images and reports. In: Computer Vision – ECCV 2022. LNCS, vol. 13686, pp. 685–701. Springer, Cham (2022). https://doi.org/10.1007/978-3-031-19809-0_39

15. Radford, A., et al.: Learning transferable visual models from natural language supervision. In: International Conference on Machine Learning, pp. 8748–8763. PMLR (2021)

16. Raghu, M., Zhang, C., Kleinberg, J., Bengio, S.: Transfusion: understanding transfer learning for medical imaging. In: Advances in Neural Information Processing Systems, vol. 32 (2019)

17. Shih, G., et al.: Augmenting the national institutes of health chest radiograph dataset with expert annotations of possible pneumonia. Radiol. Artif. Intelli. **1**(1), e180041 (2019). https://doi.org/10.1148/ryai.2019180041. pMID: 33937785

18. Singhal, K., et al.: Large language models encode clinical knowledge. arXiv preprint arXiv:2212.13138 (2022)

19. Smit, A., Jain, S., Rajpurkar, P., Pareek, A., Ng, A.Y., Lungren, M.P.: CheXbert: combining automatic labelers and expert annotations for accurate radiology report labeling using BERT. arXiv preprint arXiv:2004.09167 (2020)

20. Titano, J.J., et al.: Automated deep-neural-network surveillance of cranial images for acute neurologic events. Nat. Med. **24**(9), 1337–1341 (2018). https://doi.org/10.1038/s41591-018-0147-y. epub 2018 Aug 13

21. Tiu, E., Talius, E., Patel, P., et al.: Expert-level detection of pathologies from unannotated chest X-ray images via self-supervised learning. Nat. Biomed. Eng. (2022). https://doi.org/10.1038/s41551-022-00936-9

22. Wang, F., Zhou, Y., Wang, S., Vardhanabhuti, V., Yu, L.: Multi-granularity cross-modal alignment for generalized medical visual representation learning. In: Advances in Neural Information Processing Systems, vol. 35, pp. 33536–33549 (2022)

23. Wang, X., Peng, Y., Lu, L., Lu, Z., Bagheri, M., Summers, R.M.: ChestX-ray8: hospital-scale chest X-ray database and benchmarks on weakly-supervised classification and localization of common thorax diseases. In: Proceedings of the IEEE Conference on Computer Vision and Pattern Recognition, pp. 2097–2106 (2017)

24. Wang, Z., Wu, Z., Agarwal, D., Sun, J.: MedCLIP: contrastive learning from unpaired medical images and text. arXiv preprint arXiv:2210.10163 (2022)

25. Wu, C., Zhang, X., Zhang, Y., Wang, Y., Xie, W.: MedKLIP: medical knowledge enhanced language-image pre-training in radiology. arXiv preprint arXiv:2301.02228 (2023)

26. Xie, Y., et al.: Knowledge-based collaborative deep learning for benign-malignant lung nodule classification on chest CT. IEEE Trans. Med. Imaging **38**(4), 991–1004 (2019). https://doi.org/10.1109/TMI.2018.2876510

27. Zhang, Y., Jiang, H., Miura, Y., Manning, C.D., Langlotz, C.P.: Contrastive learning of medical visual representations from paired images and text. In: Machine Learning for Healthcare Conference, pp. 2–25. PMLR (2022)

Evaluating How Explainable AI Is Perceived in the Medical Domain: A Human-Centered Quantitative Study of XAI in Chest X-Ray Diagnostics

Gizem Karagoz[✉], Geert van Kollenburg, Tanir Ozcelebi, and Nirvana Meratnia

Eindhoven University of Technology, Eindhoven, The Netherlands
{g.karagoz,g.h.v.kollenburg,t.ozcelebi,n.meratnia}@tue.nl

Abstract. The crucial role of Explainable Artificial Intelligence (XAI) in healthcare is underscored by the need for both accurate diagnosis and transparency of decision making to improve trust in the decisions on the one hand and to facilitate its adoption by medical professionals on the other hand. In this paper, We present results of a quantitative user study to evaluate how widely used XAI methods are perceived by medical experts. For doing so, we utilize two prominent post-hoc model-agnostic XAI methods, i.e., Local Interpretable Model-agnostic Explanations (LIME) and Shapley Additive explanations (SHAP). For this study, a considerable cohort of 97 medical experts was recruited to investigate whether these XAI methods assist the medical experts in their diagnosis on Chest X-ray scans. We designed an evaluation framework to investigate diagnosis accuracy, trust change, coherence with expert reasoning, and confidence differences before and after seeing provided explanations of XAI. This large-scale study showed that both XAI methods improve scores on indicative explanations. The overall change in trust was not significantly different across LIME and SHAP, indicating that, there are other factors for trust enhancement in AI diagnostics beyond providing explanations. This work has proposed a robust, human-centered benchmark, supporting the research and development of interpretable, reliable, and clinically-aligned AI tools, and directing the future of AI in high-stakes healthcare applications towards enhanced transparency and accountability.

Keywords: Explainable AI · XAI Evaluation · Human-Centered Evaluation · XAI in Healthcare · Medical Imaging

1 Introduction

Artificial Intelligence (AI) has emerged as a cornerstone of the digital era, seamlessly integrating into both high-stakes domains and mundane daily life tasks. In non-critical applications, AI augments everyday activities often functioning without significant questioning of its decision-making processes. However, in high-stakes fields, the decision-making process becomes a matter of profound concern. The "black-box" nature of AI models, particularly in sensitive areas such as healthcare, finance, and

© The Author(s), under exclusive license to Springer Nature Switzerland AG 2024
H. Chen et al. (Eds.): TAI4H 2024, LNCS 14812, pp. 92–108, 2024.
https://doi.org/10.1007/978-3-031-67751-9_8

autonomous systems, poses a significant barrier to their broader adoption and trustworthiness [23].

Applicability of AI in healthcare has a wide variety [1] and is grouped under different application areas such as diagnosis and prognosis [13, 18, 28], drug discovery [6, 14], population health [26, 29] and so on. With the help of AI, many healthcare problems can have more accurate and faster solutions [20]. However, the most challenging obstacle between AI developments and the healthcare decision-making process is the lack of interpretability and transparency. Explainable AI (XAI) has emerged as the pivotal innovation addressing this demand, ensuring that AI decisions in medical diagnostics are not only accurate and fast but also interpretable for its human users [4].

In the emerging field of XAI, one of the most significant hurdles is developing and agreeing upon evaluation metrics that adequately measure the utility of an explanation. The effectiveness of XAI is often diminished by the inherently subjective nature of explanations; what may be clear and intuitive to one user or category of users may be opaque to another. This variability underscores the complexity of designing XAI systems that can meet the diverse needs and expectations of healthcare professionals. [19] identified that 33% of the XAI papers rely on anecdotal evidence for evaluating explanations, while only 22% of the papers conducted human-centered evaluations and approximately 5% involved domain experts in their evaluation process. Within this huge gap in the literature on the evaluation of XAI, the application-grounded evaluation which covers real human participants and real tasks of healthcare, requires to have diverse perspectives and expertise levels when assessing the quality of explanations [8]. Furthermore, the complexity of medical data often necessitates explanations that balance simplicity with the depth of details necessary for expert-level decision-making [25]. To bridge this gap, a human-centered approach for the evaluation of XAI becomes paramount.

In this paper, an application-grounded domain experts-centered approach is employed for the evaluation of state-of-the-art post-hoc model agnostic XAI methods, namely, Local Interpretable Model Agnostic Explanations (LIME) and Shapley Additive Explanations (SHAP) [16, 21]. The motivation behind the selection of these XAI methods is to provide a comprehensive evaluation and thereby establish a benchmark for the most widely used methods in healthcare [2, 15, 24]. By integrating quantitative user study evaluation methods, the study aims to uncover insights into the practical utility of XAI, navigating the intricate landscape where human intuition meets algorithmic precision.

Our contributions can be summarized as follows:

- Conducting a quantitative user study with 97 medical experts (significantly exceeding previously reported studies) divided into three groups to assess medical experts' trust, confidence, and agreement on the AI explanations for the diagnosis based on chest X-ray scans.
- Presenting key insights into medical diagnostic decision-making, showing the significant impact of AI explanations on diagnostic accuracy, trust, and professional confidence. This underscores the importance of XAI in enhancing the integration of AI within medical decision-making.

– Establishing a comprehensive baseline for future studies in healthcare by providing the initial trust levels of 97 medical experts against AI decisions, with respect to demographic distributions. This contribution is pivotal for setting a foundational benchmark for future research in enhancing AI's reliability and acceptance in healthcare.

This paper is organized as follows. Section 2 gives an overview of the literature on explainable AI in healthcare focusing on the evaluation. Section 3 describes the methodology, including the design of the user study, data collection procedures, and the statistical analysis approach with study components, variables, and hypotheses. Section 4 presents the results of the study, analyzing the impact of XAI explanations on medical experts' trust and agreement with a detailed hypothesis test procedures and implications of the findings, providing insights into how XAI can be effectively implemented in healthcare settings. Section 5 concludes the paper with a summary of the key findings, contributions to the field of human-centered AI, and suggestions for future research directions.

2   Related Work

[22] provides a comprehensive overview of applied user studies in the explainable AI's evaluation. Four core aspects, i.e., trust, understandability, usability, and human-AI collaboration performance were utilized as quality metrics for explanations. In a systematic literature review on human-centered evaluation for medical image explanations [5], 68 XAI research on medical images containing different medical image modalities and tasks (classification, segmentation, anomaly detection, and so on) were examined and the number of medical expert recruited for the evaluation of these image on transparency and interpretability varied between 1 to 30. Among these examined studies, only half of them considered medical experts as end users of the XAI systems which enlightened the lack of clinical prior knowledge validation in the serious amount of studies.

[12] proposed an evaluation chain for decision support systems in healthcare. A set of interviews was suggested to evaluate the system and the explanations regularly provided for different aspects such as business, software development, and clinical validation. Authors recruited one medical expert/annotator for the correctness of the explanations. They stated that XAI was required for all aspects since healthcare is a high-risk field for decision making and explanations alleviate the high responsibility of these vital decisions. [3] investigated the use of deep learning for enhancing medical image retrieval systems in decision-making. The aim was to identify pathologists' needs for similar image retrieval using a deep learning algorithm and to develop tools enabling users to refine search results in real-time. For validation, 12 pathologists were recruited, finding that the refinement tools not only improved the diagnostic utility of retrieved images and users' trust in the algorithm but also encouraged the adoption of new strategies to understand the algorithm. These findings suggest that such tools can significantly contribute to more effective human-AI collaborative systems in expert decision-making contexts.

With a cohort of 97 medical professionals, our study distinguishes itself from the state of the art by both its scale and depth, surpassing the existing literature where human evaluations of XAI is often engaged with considerably smaller sample sizes. This amplifies the statistical power of our findings and provides a rich, yet diverse set of insights that deepen our understanding of how XAI is perceived by practitioners.

3   User Study Design and Methodology

Our user study focuses on diagnostics using chest X-ray scans with the help of human subjects. To ensure compliance with ethical standards for research involving human participants and the General Data Protection Regulation (GDPR), an approval was obtained from the Ethical Review Board of the Eindhoven University of Technology.

3.1   XAI Methods

In the pursuit of making complex deep learning models explainable, two prominent methods have emerged, i.e., LIME and SHAP. Both methods aim to provide insights into the decision-making process of complex models, to approximate the actual decision function of the complex predictive model whether by providing a surrogate model or using other optimization approaches to provide a global interpretation of feature contributions to the overall prediction. Both of them can be applied to any predictive model (model-agnostic) after the training and validation phases are completed (post-hoc).

LIME is a technique to explain the predictions of any classifier in an interpretable and faithful manner, by approximating predictive model locally with an interpretable model [21]. The technique involves perturbing the input data and observing the corresponding changes in the black-box model's predictions. A new interpretable model is then trained on this new dataset, with the weights determined by the proximity of the perturbed instances to the instance of interest. The general formula representing LIME is given by Eq. 1 where $\xi(x)$ is the local surrogate explanation model for instance x, f is the predictive model to explain, G is the class of interpretable models, $\mathcal{L}$ is a measure of fidelity, π_x defines the locality around x, and $\Omega(g)$ is the complexity of the model g.

$$\xi(x) = \arg\min_{g \in G} \mathcal{L}(f, g, \pi_x) + \Omega(g) \tag{1}$$

SHAP leverages the concept of Shapley values from cooperative game theory to attribute the prediction output of a model to its input features [16]. It offers a unified measure of feature importance that is both consistent and locally accurate. SHAP values explain the prediction of an instance by computing the contribution of each feature to the difference between the actual prediction and the dataset's average prediction. The SHAP value for a feature j can be mathematically represented as:

$$\phi_j = \sum_{S \subseteq N \setminus \{j\}} \frac{|S|!(|N| - |S| - 1)!}{|N|!} \left[f(S \cup \{j\}) - f(S) \right] \tag{2}$$

where N is the set of all features, S is a subset of features without j, f is the model function, and ϕ_j is the SHAP value for feature j.

3.2 Pre-study Design and Participant Recruitment

Initially, the necessary groups of participants were identified, i.e., one group for the evaluation of LIME explanations, another for SHAP, and a control group without any explanations.

Subsequently, a power analysis was conducted to determine the minimum number of participants required in each group to achieve statistically significant results, with an alpha (Type-I error rate) set at 0.05. This analysis indicated that a minimum of 30 participants per group would be necessary to ensure the study's statistical power (80%) and reliability of findings, aligning with Cohen's recommendations for power analysis in behavioral science research [7].

With the ethical board's approval and completion of the power analysis, the recruitment process was initiated. The inclusion criterion was defined as being a medical expert, including medical students in their 4th to 6th years of study, general practitioners, specialty trainees, and specialists. This diverse participant base was selected to capture a broad range of insights and experiences, reflecting the varied levels of exposure and reliance on diagnostic support tools across different stages of medical training and practice.

The recruitment and data collection process were conducted in two phases to ensure participant privacy and data integrity. Initially, a recruitment form containing a consent form (detailed in the appendix) was disseminated to volunteer participants. Upon obtaining informed consent, participants were randomly assigned to one of the study groups. To further safeguard participants' privacy and to maintain the anonymity of responses, the second part of the survey was designed to be anonymous.

We addressed the prevalent biases and concerns among medical experts regarding the integration of AI in healthcare decision-making. We introduced a novel scenario involving a fictional medical student named *"A.I. Wannabee"*. This character was conceptualized as an AI entity aspiring to learn accurate diagnostic practices under the guidance of medical professionals, with the ultimate goal of serving as a supportive tool. This scenario was crafted not only to personify the AI but also to frame it as a collaborative and learning entity, rather than a competitor or a judge of medical experts'

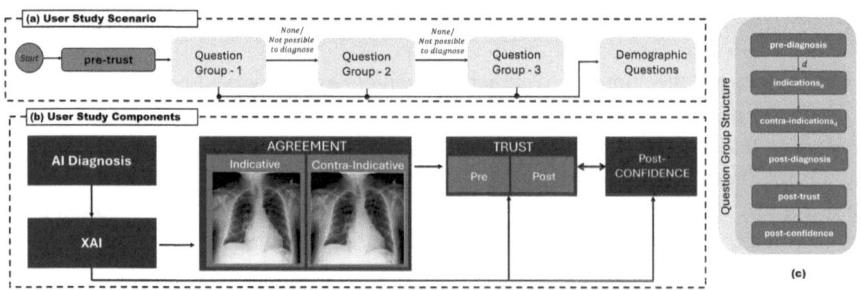

Fig. 1. Detailed User study scenario with the user study flow (a), study components with relationships between them(b), and question group structure (c).

expertise. By presenting XAI in this manner, we aimed to facilitate a more receptive and collaborative interaction between the medical experts and the AI system.

3.3 Study Components

The key components of our study include the predictive model as the base model for making diagnostic judgments on chest X-rays scans, XAI methods as techniques used to interpret how the predictive model arrives at its decision. Figure 1(b) represents user study components and the connections between them. XAI serving as a bridge to interpret AI judgments. XAI's role is critical in enhancing medical experts' agreement, trust, and confidence towards AI diagnostics. Moreover, both agreement with AI's reasoning and confidence in its judgments contribute to trust in AI, illustrating a complex interplay between understanding, confidence, and trust in AI-assisted healthcare. Additionally, dependent and independent variables in the study were defined. The independent variable in our study is the XAI method which constructs groups i.e., SHAP, LIME, Control. The dependent variables include:

- $TRUST_{pre}$: Trust in AI before the experiment begins.
- $TRUST_{post}$: Trust in AI after seeing the explanations.
- $AGREE_{ind}$: Agreement on indicative regions as identified by the AI.
- $AGREE_{cont}$: Agreement on contra-indicative regions as identified by the AI.
- $CONFIDENCE_{post}$: Confidence on the diagnosis after seeing explanations.

$TRUST_{pre}$ and $TRUST_{post}$ are measured using a 10-point Likert scale where 1 indicates "No trust at all" and 10 indicates "Complete trust". $AGREE_{ind}$ and $AGREE_{cont}$ are also measured by a 10-point Likert scale where 1 indicates "No overlap" and 10 indicates "Complete overlap" with their diagnostic indications and contra-indications respectively. Participants answered this question after they diagnosed the sample or in other words, after they generated their own reasoning. Therefore, they can compare the reasoning of AI and their own reasoning on diagnostics. Finally, the $CONFIDENCE_{post}$ is assessed by asking participants to reflect on the impact of the AI's reasoning on their judgement of the initial diagnosis. (See Appendix A for detailed questions.)

3.4 Hypotheses

To comprehensively evaluate the impact of XAI on the LIME and SHAP groups, against a control group (without explanations), we identified our hypotheses to include comparisons across all three groups. This approach allows us to examine the effects of explanations on medical experts' indicative/contra-indicative diagnostic agreement and trust in AI judgments separately. The hypotheses are listed below. μ represents the average level of dependent variable within each group. S refers to the SHAP group, L refers to the LIME group and C refers to the Control group in the equations.

- **Hypothesis 1 (H_1):** The XAI methods significantly affect the level of indicative agreement between the medical experts and the AI's diagnostic reasoning.

$$H_{1,0} : \mu_{AGREE_{ind,S}} = \mu_{AGREE_{ind,L}} = \mu_{AGREE_{ind,C}}$$
$$H_{1,1} : \exists \mu_{AGREE_{ind,i}} \neq \mu_{AGREE_{ind,j}} \mid i \neq j; \, i,j \in \{S,L,C\}$$

– **Hypothesis 2** (H_2)**:** The XAI methods significantly affect the level of contra-indicative agreement between the medical experts and the AI's diagnostic reasoning.

$$H_{2,0} : \mu_{AGREE_{cont,S}} = \mu_{AGREE_{cont,L}} = \mu_{AGREE_{cont,C}}$$
$$H_{2,1} : \exists \mu_{AGREE_{cont,i}} \neq \mu_{AGREE_{cont,j}} \,|\, i \neq j; \, i,j \in \{S, L, C\}$$

– **Hypothesis 3** (H_3)**:** The XAI methods significantly influence the change in TRUST (Δ_{TRUST}) towards the AI's diagnostic capabilities.

$$H_{3,0} : \mu_{\Delta_{TRUST},S} = \mu_{\Delta_{TRUST},L} = \mu_{\Delta_{TRUST},C}$$
$$H_{3,1} : \exists \mu_{\Delta_{TRUST},i} \neq \mu_{\Delta_{TRUST},j} \,|\, i \neq j; \, i,j \in \{S, L, C\}$$

– **Hypothesis 4** (H_4)**:** The level of agreement between the medical experts and the AI's diagnostic reasoning significantly affect the change in trust (Δ_{TRUST}) towards the AI.

$$H_{4,0} : \forall Group \in \{S, L, C\}, \, \beta_{AGREE \to \Delta_{TRUST}, Group} = 0$$
$$H_{4,1} : \exists Group \,|\, \beta_{AGREE \to \Delta_{TRUST}, Group} \neq 0$$

Incorporating the control group allows us to not only determine the effectiveness of each XAI method but also understand how the provision of any explanation (versus none) influences medical experts' agreement with AI judgments and their trust in AI's diagnostic capabilities. The null hypotheses ($H_{i,0}$) assume no difference in the testing variable change across the groups, while the alternative hypotheses ($H_{i,1}$) suggest that differences exist, warranting further investigation into the specific impacts of LIME, SHAP, and the absence of explanations.

3.5 User Study Scenario

The user study scenario as a flowchart is shown in Fig. 1(a). In our user study, participants commenced by assessing their initial trust in AI's diagnostic capabilities, recorded as the $TRUST_{pre}$ measure. Following this initial assessment, participants were presented a chest X-ray scan for diagnosis (pre-diagnosis). For each scan, participants were given the option to diagnose based on a set of possible diseases or to select "None/Not possible to diagnose" if they believed a diagnosis could not be determined or the conditions listed did not apply to the patient. In this case, participants were directed to a new X-ray scan sample, preserving the study's integrity by ensuring every participant engaged with a diagnosable case. This process allowed up to three attempts with different X-ray samples (question groups 1–3) to accommodate varying levels of diagnostic clarity. Each X-ray sample in each question group, was carefully selected based on similar levels of diagnostic difficulty, which was determined by the predictive accuracies and their respective ground truths which is verified by medical professionals in the dataset. This approach was designed to minimize variability in diagnostic difficulty across samples.

Considering the diagnostic decision, participants were then shown supportive (indication) and unsupportive (contra-indication) explanations of their diagnosis (d) and then asked for a rating on a scale from 1 to 10 to evaluate their agreement with the AI's reasoning for the first two groups, which contain explanations. In the control group, to maintain consistency in question distribution, participants were asked to express their opinions on the importance of using indications and contra-indications for diagnosis. Similarly, they evaluated the role of these indications and contra-indications in the diagnostic process. Subsequently, the AI's judgment (classification prediction) for the diagnosed sample was revealed, and participants were inquired if AI judgment with its provided explanation altered their initial diagnosis, constituting the post-diagnosis measure. This step was followed by queries on whether exposure to the explanations altered their confidence in making diagnoses ($CONFIDENCE_{post}$) and a final assessment of their trust in AI post-exposure ($TRUST_{post}$).

Upon completion of the diagnostic and explanatory feedback loop, participants proceeded to answer demographic questions, concluding the user study.

4 Analysis and Results

4.1 Predictive and XAI Model Setup

We utilized the publicly available Chest X-Ray dataset [27] from National Institutes of Health (NIH). The dataset comprises 112,120 frontal-view X-ray images sourced from 30,805 unique patients, making it one of the largest publicly available chest x-ray datasets. Each image within the dataset is labeled as belonging to one or more of 14 available classes, including 13 disease classes and another class called "No Finding" for images that do not exhibit any of the 13 listed pathologies. The diseases represented in the dataset span a range of common to rare chest-related conditions, such as Atelectasis, Cardiomegaly, Effusion, Infiltration, Mass, Nodule, Pneumonia, Pneumothorax, Consolidation, Edema, Emphysema, Fibrosis, Pleural Thickening and Hernia.

For the predictive analysis in our study, we employed DenseNet121, a convolutional neural network known for its efficiency and accuracy in image classification tasks, including medical imaging. We utilized TensorFlow's implementation of DenseNet121, leveraging a pre-trained model on the NIH Chest X-Ray dataset to enhance its predictive performance. The model's performance was assessed using the accuracy and Area Under the Curve (AUC) metrics for each of the labels, yielding a mean AUC of 0.80121 and a mean accuracy of 0.70983. These metrics indicate that the model achieved an acceptable level of predictive performance, making it a suitable basis for explanation via XAI methods.

For SHAP, the Kernel SHAP implementation was used. Both XAI methods operate as surrogate models that approximate the predictions of the original complex black-box model; therefore, to ensure a robust analysis and facilitate a fair comparison between LIME and SHAP, we set the number of samples parameter to 1000 for both methods. This choice not only aligns the experimental conditions for each method but also provides a comprehensive foundation for comparing their efficiency in making the model's decisions interpretable.

4.2 User Study Evaluation Analysis

Demographic Distribution. In our study, demographic analysis across the three groups, i.e., LIME, SHAP, and Control, reveals a diverse participation encompassing various ages, genders, countries of residence, proficiency levels, and academic titles (see Fig. 2). Notably, the age distribution indicated a predominant participation from the 25–34 and 35–44 age ranges, reflecting a younger demographic of medical professionals and students engaged in the study. Gender representation showed a balanced mix, with a slight predominance of male participants, consistent across all groups. The majority of participants identified as either specialty trainees or specialists, demonstrating the study's appeal to those with substantial medical education and experience. The country of residence was predominantly Turkey, followed by the Netherlands, indicating a geographic diversity in the study's reach. Academic titles varied, with a significant number of participants holding advanced degrees, underscoring the study's relevance to a highly educated audience in the medical field. Moreover, our demographic analysis extends to the diverse specialties and corresponding years of experience of the medical experts participating in the study.

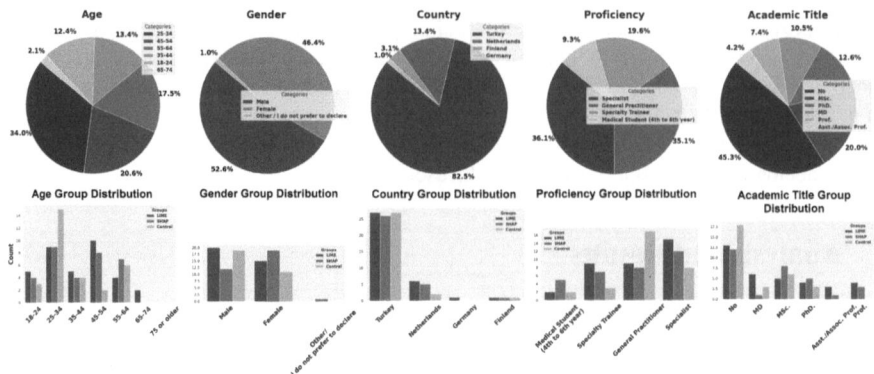

Fig. 2. Demographic statistics and group distribution

Across all groups, we observed a wide range of specialties, from Lung Diseases, Gynecology and Obstetrics, to Radiology and Nuclear Medicine, indicating the study's comprehensive appeal to medical professionals across various fields. This diversity in specialties and experience enriches the study's findings by incorporating a broad spectrum of medical expertise and perspectives.

Before conducting the hypothesis tests, a normality check was applied on all variables using the Kolmogorov-Smirnov test [17]. This step was crucial to ensure that the assumptions underlying parametric statistical tests, such as analysis of variance (ANOVA) [10] and t-tests, were met. The paired t-test checks the mean differences of the same group before and after an intervention, producing a t-score that measures the relative change and a p-value indicating the chance of observing such a difference if there was no actual effect. ANOVA, on the other hand, compares means across multiple groups, yielding an f-statistic to indicate the variance ratio between and within groups and a p-value to assess the likelihood of these group differences occurring by chance [9]. The results of the normality check confirmed that all variable distributions conformed to a normal distribution, satisfying the necessary precondition for the subsequent analyses. Together, the successful normality check and the confirmation of sufficient study power laid a solid foundation for the integrity and credibility of our hypothesis testing process.

Additionally, the distribution of initial trust scores is calculated among all of the groups to assign the trust perspective of medical experts for AI involvement in medical diagnosis. The mean value of $TRUST_{pre}$ is recorded as 5.96 and the most dense bins are 5, 6, 7, and 8. Therefore, a moderate trust was examined.

Agreement Analysis (H_1, H_2)
Indicative Agreement: To examine differences in average agreement scores across three groups the one-way ANOVA was conducted to test H_1. The results of indications yielded an f-statistic (i.e., a ratio of variances) of 5.608 and a p-value of 0.005, indicating that there are statistically significant differences in agreement scores among the groups. This result led us to reject the null hypothesis ($H_{1,0}$), affirming that not all group means are equal and that the type of explanation (or the absence thereof) impacts agreement levels. The Control group showed a mean indicative agreement score of 6.167, which serves as the baseline, reflecting the inherent value medical experts place on indications and contra-indications in their diagnostic reasoning. LIME showed a mean indication score of 6.030, slightly below the Control group but still represents a high level of indicative agreement. Finally, SHAP presented a relatively lower score of 4.344, suggesting that SHAP explanations may not align as closely with medical experts' reasoning.

To interpret more on ANOVA results, the Tukey HSD (honestly significant difference) post-hoc analysis was implemented, which revealed that LIME maintained agreement levels near this baseline (mean difference from Control group = 0.1364, p = 0.9728, not significant), suggesting that LIME effectively mirrors the diagnostic considerations valued by medical experts. Conversely, SHAP significantly deviates from the baseline (mean difference from Control group is 1.8229, p = 0.0105), highlighting a disconnect between SHAP explanations and the experts' diagnostic values.

Contra-indicative Agreement: Similarly, one-way ANOVA and subsequently Tukey HSD were applied to test $AGREE_{cont}$ changes of the three groups. The ANOVA yielded an f-statistic of 5.163 with a p-value of 0.007, suggesting significant differences in the emphasis on contra-indications across the groups. This result led to the rejection of the null hypothesis, affirming that the type of explanation or absence thereof impacts the perceived value of contra-indications.

Control group with a mean of 5.933, established the baseline, indicating a strong inherent emphasis on contra-indications in the diagnostic process. LIME had a mean of 4.303, suggesting that while LIME explanations are valued, they may not fully capture the importance of contra-indications as perceived in the control group. SHAP, with a mean of 4.031, showed a similar trend to LIME but is slightly lower, indicating a greater difference in the perceived value of contra-indications provided by SHAP explanations.

Tukey HSD post-hoc analysis results were as follows, for LIME-SHAP: mean diff = -0.2718, p = 0.9005, LIME-Control group: mean diff = 1.6303, p = 0.031 and finally for SHAP-Control group: mean diff = 1.9021, p = 0.0102. The findings, especially the significant differences observed between both explanation groups and the control group, illuminate the critical role of contra-indications in diagnostic decision-making. The higher emphasis on contra-indications in the control group suggests that, while LIME and SHAP explanations provide valuable insights, they may not fully align with the inherent diagnostic considerations of medical experts, particularly regarding contra-indicative agreement.

Trust Analysis (H_3). To assess if there's a statistically significant change in level of trust between pre to post-intervention, the paired t-test over $TRUST_{pre}$ and $TRUST_{post}$ was applied for each group. Paired t-test results within each group revealed that neither the provision of LIME/SHAP explanations nor the absence of such explanations in the control group significantly altered the trust that medical experts placed in AI's diagnostic capabilities. The t-test results demonstrate stability in trust levels, suggesting that, within the scope of this study, providing LIME/SHAP explanations did not markedly influence trust in AI diagnostics.

To analyze the trust differences in between the groups deeper, we calculated a derived variable, i.e., Δ_{TRUST}, basically from the difference of $TRUST_{post}$ and $TRUST_{pre}$. Then, we applied ANOVA to compare trust differences between the groups. Tukey HDS post-hoc test was applied subsequently. However, pairwise comparisons between groups resulted in non-significant differences (p-values > 0.05), further supporting the conclusion that the type of explanation provided (or its absence) does not significantly influence the change in trust levels for AI judgment among participants.

Considering the flow of user study, we alternatively focused on question group-based analysis which considers whether participants diagnosed the first, second, or third question groups mentioned in Sect. 3. It was observed that a significant majority of participants in the LIME, SHAP, and Control group responded to the first question group, with respective counts of 30/35 in LIME, 20/32 in SHAP, and 29/30 in the Control group. Consequently, we explored the impact of the question group that participants diagnosed on the independent variables. This analysis, which considered the specific question group each participant responded to, yielded results on Δ_{TRUST} that were consistent with those obtained from the overall analysis, which did not differ per question group. The post-confidence question was formulated to serve as a support variable to trust differences among the groups. Participants were asked to indicate whether their confidence in their diagnosis was positively affected, negatively affected, or remained unchanged after viewing explanations. The majority of participants indicated that their diagnostic confidence was not affected by the explanations. To assess the variations in confidence levels after seeing explanations, ANOVA and Tukey's HSD tests were employed. Similar to the findings on Δ_{TRUST}, the analysis revealed no significant differences in confidence changes across the groups, indicating a uniform effect of explanations on participants' confidence in their diagnostic decisions.

Agreement Trust Relationship Analysis (H_4). To evaluate the impact of agreement with AI's diagnostic reasoning on changes in trust towards explanations (Δ_{TRUST}), Pearson correlation analysis was employed, reflecting the continuous nature of the variables involved [11]. This analytical method is designed to check whether there is a significant correlation between $AGREE$ and Δ_{TRUST}, thus challenging the null hypothesis ($H_{4,0}$) which proposes no relationship (correlation coefficient (β) = 0) between these dimensions.

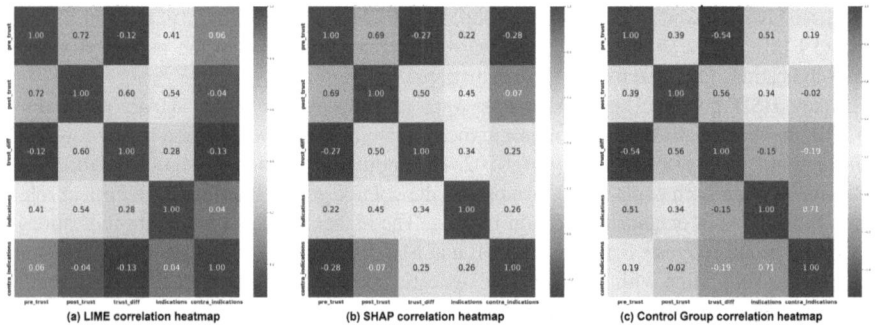

Fig. 3. Correlation heatmaps for pre-trust, post-trust, indicative and contra-indicative agreement over groups

The correlation analysis (see Fig. 3) for LIME shows a significant relationship between $AGREE_{ind}$ and both $TRUST_{pre}$ and $TRUST_{post}$, with correlation coefficients (β) of 0.412 and 0.536, respectively. This suggests that $AGREE_{ind}$ of LIME positively influenced Δ_{TRUST} in AI's diagnostic. However, the Δ_{TRUST} exhibited a slight negative correlation with $TRUST_{pre}$ (r = −0.123), indicating that an initial high level of trust may not lead to significant Δ_{TRUST} after seeing the LIME explanations.

In SHAP, a positive correlation with $TRUST_{post}$ (r = 0.450) is observed, which is weaker than in LIME, suggesting that SHAP explanations also contributed to increasing trust through indications but possibly to a lesser extent than LIME. However, a moderate and slight negative correlation is identified between $AGREE_{cont}$ and trust levels respectively $TRUST_{pre}$ and $TRUST_{post}$. Notably, Δ_{TRUST} positively correlated with $TRUST_{post}$ (r = 0.504), indicating that experiencing SHAP explanations can lead to an increase in trust levels.

The Control group, showed the weakest correlations between $AGREE_{ind,cont}$ and Δ_{TRUST}. The slight positive correlation between $TRUST_{pre}$ and $TRUST_{post}$ (r = 0.393) suggests a baseline level of trust stability unaffected by explanatory interventions, underscoring the importance of explanations in influencing trust dynamics. The strong negative correlation between $TRUST_{pre}$ and Δ_{TRUST} (r = −0.544) further illustrates the challenge in altering trust without the aid of explanations. As expected, the positive correlation between $AGREE_{ind}$

and $AGREE_{cont}$ (r = 0.706) is quite high which indicates both having indications and contra-indications have similar importance in diagnosis.

Diagnostic Accuracy-Based Analysis. Additionally, the diagnostic accuracy and its effect were tested on Δ_{TRUST}, $AGREE_{ind}$, and $AGREE_{cont}$. It is examined under two conditions such as when participants' diagnoses are the same as the ground truth and different from the ground truth. Under these conditions, the paired t-test is applied to check $TRUST_{pre}$ and $TRUST_{post}$ differences within the groups. Similar to the other hypothesis tests, Δ_{TRUST} between the groups is checked with the ANOVA and Tukey's HDS. Mean differences of Δ_{TRUST} and $AGREE$ are shown in Table 1.

Table 1. Means for independent variables under diagnostic accuracy condition and overall

	Overall			Diagnosis = GT		
	Δ_{TRUST}	$AGREE_{ind}$	$AGREE_{cont}$	Δ_{TRUST}	$AGREE_{ind}$	$AGREE_{cont}$
LIME	0.200	6.030	4.303	0.833	7.042	4.000
SHAP	0.125	4.344	4.031	−0.267	4.133	3.600
Control	0.500	6.167	5.933	1.000	6.045	5.773

The ANOVA results revealed significant differences in Δ_{TRUST} across the groups (f-statistic = 5.803, p = 0.005), indicating that diagnostic accuracy significantly affects trust alteration. Specifically, control group exhibited the highest mean increase in trust (1.000), followed by LIME with a mean increase of 0.833, and SHAP displayed a mean decrease in trust (−0.267). The Tukey HSD test highlighted a significant decrease in trust between LIME and SHAP (p = 0.0168), indicating that LIME explanations had gained trust against the SHAP explanations when the diagnosis is equal to the ground truth. In overall analysis, Δ_{TRUST} in LIME recorded as 0.2 however, when participant performed accurate diagnosis, Δ_{TRUST} had increased to 0.833. It is not the same for the SHAP explanation since it degraded from 0.125 to −0.267 indicating Δ_{TRUST} had decreased on the ground-truth accurate diagnosis.

Indications (ANOVA: f-statistic = 11.259, p-value = 0.0001) and contra-indications(ANOVA: f-statistic = 4.998, p-value = 0.010) were also analyzed with the same methodology and provided significant distribution differences among groups. These findings are similar to the overall analysis results. For LIME, $AGREE_{ind}$ was increased from 6.03 to 7.04. On the other hand, $AGREE_{cont}$ was decreased more when diagnostic accuracy was considered, which stresses the consistency of the findings.

5   Conclusion

In an era where AI's potential in healthcare hinges on the transparency of its decision-making processes, our study confronts a pivotal question: *"How do medical experts trust and perceive the AI explanations in critical diagnostic scenarios?"*. Through engaging 97 medical experts, we aimed to understand the interplay between trust, interpretability, and confidence with AI diagnostics. Utilizing the 'A.I. Wannabee' scenario, we mitigated biases and examined the effects of LIME and SHAP explanations versus a control group without explanations on expert agreement levels. Our findings indicate that while LIME closely aligns with expert reasoning through indications, SHAP explanations show slight misalignment on indications. However, in presenting contra-indicative explanations, both methods did not provide high performance.

Within the analysis of Δ_{TRUST} between before and after seeing explanations did not reveal significant changes for any groups. The lack of significant differences in trust change across groups suggests that enhancing trust in AI diagnostics may require more than just providing explanations. This aligns with the broader narrative that trust in AI systems is built on multiple dimensions, including but not limited to the quality and nature of explanations. Future research should explore other factors that may influence trust in AI diagnostics, such as user experience, the accuracy of AI predictions, and the integration of AI systems into clinical workflows.

The correlation analysis revealed that trust in AI diagnostics is significantly influenced by the quality of explanations, with LIME showing a stronger positive impact on trust through support-ive evidence compared to SHAP. The minimal correlation observed in the control group under-scores the vital role of explanations in building trust, emphasizing the importance of a balanced approach that incorporates both supportive and unsupportive evidence to foster trust in AI diag-nostics effectively. Additionally, The initial trust values, gathered from a diverse group of 97 medical experts, provide a baseline measurement of trust towards AI judgment in the healthcare domain. The mean initial trust is recorded as 6 over 10 which indicates the trust for AI involve-ment in diagnosis is moderate.

Our study's contributions are substantial, setting a new benchmark for human-centered eval-uations of XAI in healthcare. The extensive participant base allows us to offer one of the most comprehensive evaluations of post-hoc XAI methods to date, serving as a reliable reference for future research and development in the field. By presenting empirical evidence from a signifi-cant sample of medical experts, our findings not only enrich the academic discourse on XAI but also have the potential to inform policies and best practices for integrating AI into healthcare diagnostics.

Acknowledgments. We would like to extend our gratitude to all participants, without whom this study could not be carried out. This work has been done in the context of the ITEA 3, Privacy preserving cross-organizational data analysis in the healthcare sector (Secure e-Health) project.

A   Appendix

In the introduction to the question group, Fig. 4 is provided to participants to explain what they should expect from questions also to be able to answer the $TRUST_{pre}$ question within the context . Then, $TRUST_{pre}$ as in Fig. 5 (left), pre-diagnosis as in Fig. 6 (left), $AGREE_{ind}$ and $AGREE_{cont}$ questions are provided regarding the diagnosis made by the participants as in Fig. 7. After explanations as supportive/indicative and unsupportive/contra-indicative are pro-vided to the participant post-diagnosis question is shown in Fig. 6 (right). Before the demographic part, $TRUST_{post}$ is asked as in Fig. 5(right) to be ranked by the participants.

To understand the baseline trust levels of medical experts in AI diagnostics, we analyzed the $TRUST_{pre}$ distribution across various demographic variables, including age, country, pro-ficiency, and academic title (See Fig. 8). We found that ages between 35–44 and 65–74 have relatively higher initial trust levels in AI diagnostic abilities. Regarding the impact of academic title on trust levels, medical experts with MSc and PhD degrees demonstrate higher trust levels compared to other groups, with the lowest trust levels recorded among participants without an academic title. Surprisingly, proficiency appears to have no significant effect on trust levels in AI diagnostics, as all proficiency groups exhibit similar trust levels.

Question Group Introduction

In this question group, I will share an X-ray scan and corresponding diagnosis predictions made by me. Also, you will see reasoning behind my decisions. All of the samples will be chest X-Ray scans and the purpose is to diagnose them with one of the following medical conditions:

- Cardiomegaly, Emphysema, Effusion, Hernia, Infiltration, Mass, Nodule, Atelectasis, Pneumothorax, Pleural Thickening, Pneumonia, Fibrosis, Edema, Consolidation.

Fig. 4. Question group introduction in a non-control group.

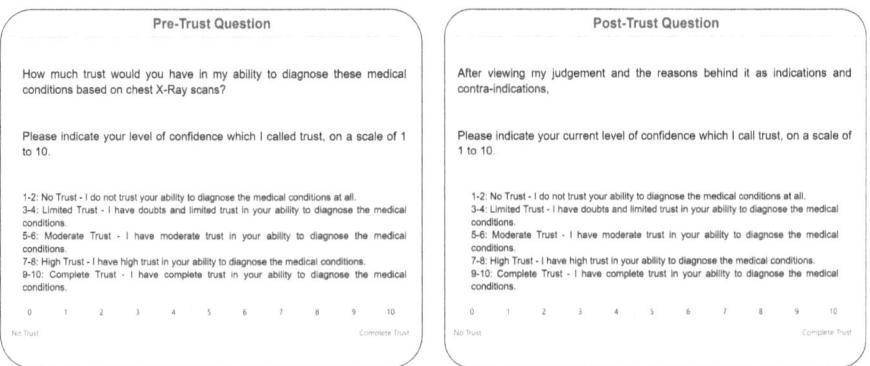

Fig. 5. A sample question for pre-trust which is asked at the beginning of the user study and post-trust indicates the trust level after seeing AI judgment with corresponding explanations of diagnosis in a non-control group.

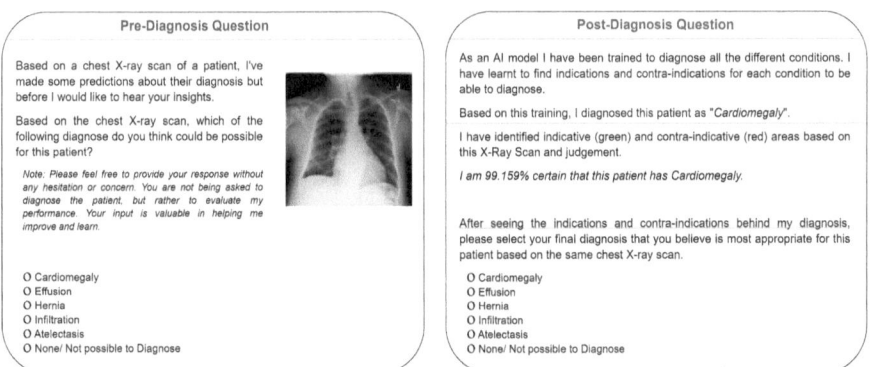

Fig. 6. A sample question for both pre-diagnosis represents the question which is the initial diagnosis before any AI judgment or explanation and the post-diagnosis indicates the diagnosis after AI judgment with corresponding explanations of diagnosis in a non-control group.

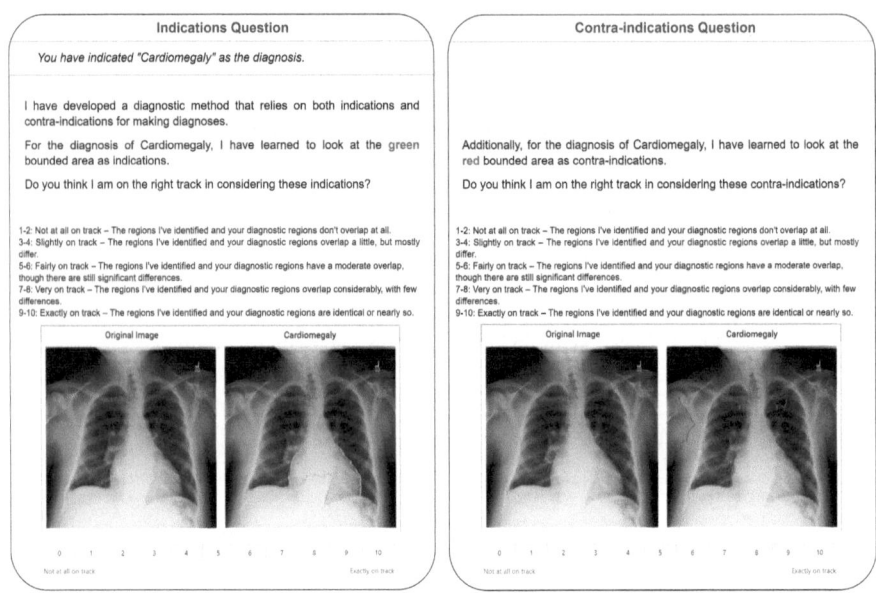

Fig. 7. Sample questions for both indications and contra-indications regarding the diagnosis provided by the participant in a non-control group.

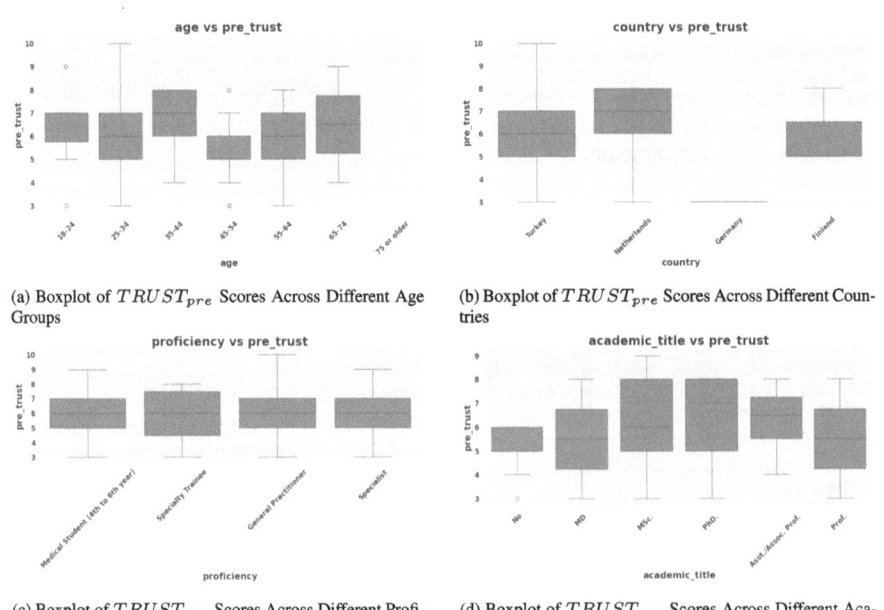

(a) Boxplot of $TRUST_{pre}$ Scores Across Different Age Groups

(b) Boxplot of $TRUST_{pre}$ Scores Across Different Countries

(c) Boxplot of $TRUST_{pre}$ Scores Across Different Proficiencies

(d) Boxplot of $TRUST_{pre}$ Scores Across Different Academic Titles Status

Fig. 8. Comparative Analysis of Pre-Trust Scores by Demographic and Professional Variables

References

1. Adadi, A., Berrada, M.: Explainable AI for healthcare: from black box to interpretable models. In: Bhateja, V., Satapathy, S.C., Satori, H. (eds.) Embedded Systems and Artificial Intelligence. AISC, vol. 1076, pp. 327–337. Springer, Singapore (2020). https://doi.org/10.1007/978-981-15-0947-6_31
2. Bhandari, M., Yogarajah, P., Kavitha, M.S., Condell, J.: Exploring the capabilities of a lightweight cnn model in accurately identifying renal abnormalities: Cysts, stones, and tumors, using lime and shap. Appl. Sci. **13**(5), 3125 (2023)
3. Cai, C.J., et al.: Human-centered tools for coping with imperfect algorithms during medical decision-making. In: Proceedings of the 2019 Chi Conference on Human Factors in Computing Systems, pp. 1–14 (2019)
4. Caruana, R., Lou, Y., Gehrke, J., Koch, P., Sturm, M., Elhadad, N.: Intelligible models for healthcare: Predicting pneumonia risk and hospital 30-day readmission. In: Proceedings of the 21th ACM SIGKDD International Conference on Knowledge Discovery and Data Mining, pp. 1721–1730 (2015)
5. Chen, H., Gomez, C., Huang, C.M., Unberath, M.: Explainable medical imaging AI needs human-centered design: guidelines and evidence from a systematic review. NPJ Digit. Med. **5**(1), 156 (2022)
6. Chen, Z., Liu, X., Hogan, W., Shenkman, E., Bian, J.: Applications of artificial intelligence in drug development using real-world data. Drug Discovery Today **26**(5), 1256–1264 (2021)
7. Cohen, J.: Statistical Power Analysis for the Behavioral Sciences. Academic Press (2013)
8. Doshi-Velez, F., Kim, B.: Considerations for evaluation and generalization in interpretable machine learning. In: Escalante, H.J., et al. (eds.) Explainable and Interpretable Models in Computer Vision and Machine Learning. TSSCML, pp. 3–17. Springer, Cham (2018). https://doi.org/10.1007/978-3-319-98131-4_1
9. Field, A.: Discovering statistics using IBM SPSS statistics. Sage publications limited (2024)
10. Fisher, R.A., et al.: The Design of Experiments. The Design of Experiments, 7th edn. (1960)
11. Freedman, D., Pisani, R., Purves, R.: Statistics (international student edition). Pisani, R. Purves, 4th edn. WW Norton & Company, New York (2007)
12. Gerlings, J., Jensen, M.S., Shollo, A.: Explainable AI, but explainable to whom? an exploratory case study of XAI in healthcare. In: Handbook of Artificial Intelligence in Healthcare: Vol 2: Practicalities and Prospects, pp. 169–198 (2022)
13. Ghouali, S., et al.: Artificial intelligence-based teleopthalmology application for diagnosis of diabetics retinopathy. IEEE Open J. Eng. Med. Biol. **3**, 124–133 (2022)
14. Iswarya, B., Manimekalai, K.: Drug discovery with XAI using deep learning. In: Principles and Methods of Explainable Artificial Intelligence in Healthcare, pp. 131–149. IGI Global (2022)
15. Knapič, S., Malhi, A., Saluja, R., Främling, K.: Explainable artificial intelligence for human decision support system in the medical domain. Mach. Learn. Knowl. Extract. **3**(3), 740–770 (2021)
16. Lundberg, S.M., Lee, S.I.: A unified approach to interpreting model predictions. In: Guyon, I., Luxburg, U.V., Bengio, S., Wallach, H., Fergus, R., Vishwanathan, S., Garnett, R. (eds.) Advances in Neural Information Processing Systems, vol. 30, pp. 4765–4774. Curran Associates, Inc. (2017). http://papers.nips.cc/paper/7062-a-unified-approach-to-interpreting-model-predictions.pdf
17. Massey, F.J., Jr.: The kolmogorov-smirnov test for goodness of fit. J. Am. Stat. Assoc. **46**(253), 68–78 (1951)
18. Mitsala, A., Tsalikidis, C., Pitiakoudis, M., Simopoulos, C., Tsaroucha, A.K.: Artificial intelligence in colorectal cancer screening, diagnosis and treatment. a new era. Current Oncol. **28**(3), 1581–1607 (2021)

19. Nauta, M., et al.: From anecdotal evidence to quantitative evaluation methods: a systematic review on evaluating explainable AI. ACM Comput. Surv. **55**(13s), 1–42 (2023)
20. Rajpurkar, P., et al.: ChexNet: radiologist-level pneumonia detection on chest x-rays with deep learning. arXiv preprint arXiv:1711.05225 (2017)
21. Ribeiro, M.T., Singh, S., Guestrin, C.: "why should i trust you?" explaining the predictions of any classifier. In: Proceedings of the 22nd ACM SIGKDD International Conference on Knowledge Discovery and Data Mining, pp. 1135–1144 (2016)
22. Rong, Y., et al.: Towards human-centered explainable AI: a survey of user studies for model explanations. IEEE Trans. Pattern Anal. Mach. Intell. (2023)
23. Rudin, C.: Stop explaining black box machine learning models for high stakes decisions and use interpretable models instead. Nat. Mach. Intell. **1**(5), 206–215 (2019)
24. Severn, C., Suresh, K., Görg, C., Choi, Y.S., Jain, R., Ghosh, D.: A pipeline for the implementation and visualization of explainable machine learning for medical imaging using radiomics features. Sensors **22**(14), 5205 (2022)
25. Tjoa, E., Guan, C.: A survey on explainable artificial intelligence (XAI): toward medical XAI. IEEE Trans. Neural Networks Learn. Syst. **32**(11), 4793–4813 (2020)
26. Wahl, B., Cossy-Gantner, A., Germann, S., Schwalbe, N.R.: Artificial intelligence (AI) and global health: how can AI contribute to health in resource-poor settings? BMJ Glob. Health **3**(4), e000798 (2018)
27. Wang, X., Peng, Y., Lu, L., Lu, Z., Bagheri, M., Summers, R.M.: Chestx-ray8: hospital-scale chest x-ray database and benchmarks on weakly-supervised classification and localization of common thorax diseases. In: Proceedings of the IEEE Conference on Computer Vision and Pattern Recognition, pp. 2097–2106 (2017)
28. Xu, X., et al.: A clinically applicable AI system for diagnosis of congenital heart diseases based on computed tomography images. Med. Image Anal. **90**, 102953 (2023)
29. Zhang, H., et al.: Deep learning based drug screening for novel coronavirus 2019-ncov. Interdiscip. Sci. Comput. Life Sci. **12**, 368–376 (2020)

SnapSeg: Training-Free Few-Shot Medical Image Segmentation with Segment Anything Model

Nanxi Yu[1,2], Zhiyuan Cai[1,3], Yijin Huang[1,4], and Xiaoying Tang[1,3(✉)]

[1] Southern University of Science and Technology, Shenzhen, China
tangxy@sustech.edu.cn
[2] The Hong Kong Polytechnic University, Hong Kong SAR, China
[3] Jiaxing Research Institute, Jiaxing, China
[4] The University of British Columbia, Vancouver, Canada

Abstract. In the pursuit of advancing medical diagnosis, automatic segmentation of medical images is crucial, particularly in extending medical expertise to under-resourced regions. However, collecting and annotating medical data for deep learning frameworks are both time-consuming and expensive. Few-shot learning, which leverages limited labeled data to learn new tasks, has been widely applied to medical image segmentation, offering significant advancements. Nonetheless, these methods often rely on extensive unlabeled data to acquire prior medical knowledge. We introduce **SnapSeg**, a novel few-shot segmentation framework that stands out by requiring only a minimal set of labeled images to directly tackle new segmentation tasks, thus bypassing the need for a traditional training phase. Utilizing either a single or a few labeled examples, SnapSeg extracts multi-level features from the Segment Anything Model (SAM)'s image encoder and incorporates a relative anchor algorithm for precise spatial assessment. Our method demonstrates state-of-the-art performance on the widely-used Abd-CT dataset in medical image segmentation.

Keywords: Few-shot Learning · Medical Image Segmentation · Segment Anything Model

1 Introduction

Automatic segmentation of medical images plays a crucial role in supporting medical diagnoses [31], especially by providing essential medical knowledge to areas with inadequate healthcare facilities. However, within the context of deep learning in healthcare, the acquisition and labeling of medical data are notably labor-intensive and costly [6]. Consequently, conventional deep learning methods for segmentation tasks prove impractical in such contexts. Given this backdrop, the robust generalization potential of Few-shot Learning (FSL) [18,19,22] emerges as a promising solution for healthcare initiatives, as FSL aims to create models capable of effective generalization with minimal labeled data. Although

H. Chen et al. (Eds.): TAI4H 2024, LNCS 14812, pp. 109–122, 2024.
https://doi.org/10.1007/978-3-031-67751-9_9

the primary objective of FSL is to efficiently learn new tasks with only a few labeled data, conventional FSL methods [20,25] still require substantial labeled data during their training phase, rendering them impractical for the medical image domain. To address this limitation, recent studies in the domain of few-shot segmentation within medical image domain [1,3,15,29,30] predominantly utilize metric learning augmented by self-supervised techniques, relying on substantial volumes of unlabeled data for training. [29] calculates multiple regional prototypes to reduce the effects of intra-class variability, while [1,29,30] introduce supplementary learnable modules to optimize the interaction between support and query images, thus developing more robust and comprehensive prototype representations. Despite these advancements, these methods still heavily depend on large volumes of unlabeled data during their training stages.

As shown in Fig. 1, SnapSeg comprises three modules: (1) Generation module: Leveraging the robust zero-shot generalization capabilities of the Segment Anything Model (SAM) [7,12], we evaluate the characteristics of various feature levels in SAM's image encoder. Our results show that mid-level features predominantly focus on local positioning, while high-level features increasingly concentrate on global distribution and morphology. To ensure robust and adaptable segmentation performance across diverse medical tasks, we generate multi-level features to enrich the information available for segmentation analysis and produce a set of masks for a query image as the object proposal set. (2) Multi-level similarity analyser module: We generate similarity maps to assess the relationship between the target object and the images at a fine-grained level, utilizing these maps to filter the high-related candidates from the object proposal set. (3) Dual perspective prediction module: This module analyzes correlations between the target class in the support image and the highly related candidates in the query image from both feature and spatial perspectives, finalizing the mask prediction for the query image. Furthermore, in terms of spatial consistency, medical images of the same category generally exhibit consistent morphological structures. However, discrepancies arise across different images due to their derivation from various standardized equipment, affecting the exact positioning, alignment, and orientation of the organs. To capitalize on this inherent consistency while addressing these challenges, we introduce the relative anchor algorithm in the Dual perspective prediction module, to accurately measure the spatial relationships between the target object and the images. Extensive experiments are conducted on the publicly available abdominal CT dataset [8], with results demonstrating consistent improvements over state-of-the-art FSS methods in the medical image domain.

2   Related Work

2.1   Few-Shot Learning

Few-shot Learning (FSL) aims to devise models capable of effective generalization with minimal labeled data. Contrary to the traditional machine learning paradigm that often demands extensive data, FSL seeks to emulate human-like learning, where we frequently assimilate new features based on few samples.

Existing few-shot learning methodologies can be broadly classified into: Metric learning approaches [14,18,22,27], Optimization-based approaches [2,9,28], and Transfer-learning approaches [10,11,13]. Metric learning methods employ visual encoders to map images into an embedding space and subsequently utilize a function to measure the similarity score between support samples and query samples. This similarity function can be predefined, such as the Euclidean distance [18,22], or can manifest as a learnable, more precise metric [4,14]. Optimization-based methods, e.g. MAML [2], train network parameters through a two-stage optimization, ensuring fast generalization performance for the current task with minimal gradient steps. Transfer-learning strategies, like S2M2 [13], generate samples to mitigate data scarcity and subsequently execute task-level fine-tuning to enhance performance.

2.2   Few-Shot Semantic Segmentation

Semantic segmentation could be regarded as pixel-level classification. Due to the scarcity of annotated data, Few-Shot Semantic Segmentation (FSS) has garnered significant attention in recent years. Shaban et al. [17] introduce the first method for addressing FSS, employing a novel dual-branch approach for one-shot semantic segmentation tasks. Wang et al. [24] integrate prototype alignment regularization atop a standard convolutional network, encouraging the model to learn consistent embedding prototypes for enhanced performance. Zhang et al. [26] introduce an Iterative Optimization Module (IOM) on the concatenated query and support features to refine segmentation predictions. Tian et al. [21] harness high-level features for prior mask production, and then address spatial inconsistencies by enhancing query features with the integration of support features and prior masks. Despite these advances, these studies still adhere to a fully-supervised paradigm, necessitating extensive labeled data for training. This requirement becomes particularly challenging in the medical image domain, where annotating images is not only time-consuming but also costly.

2.3   Medical Image Few-Shot Segmentation

In the domain of few-shot segmentation within medical imaging, studies such as [20,25] follow a fully-supervised paradigm, requiring extensive labeled data for consistent training and testing on the same tasks. Conversely, some works [1,3,15,29,30] predominantly employ metric learning augmented by self-supervised techniques, relying on substantial volumes of unlabeled data for both training and testing on comparable tasks. Yu et al. [25] innovatively utilize a grid approach to dissect images into patches, generating multiple local prototypes that encode spatial data details. Zhu et al. [29] calculate multiple regional prototypes to diminish the effects of intra-class variability, while [1,20,29,30] introduce supplementary learnable modules to optimize the interaction between support and query images, thereby cultivating more robust and comprehensive prototype representations. Despite these advancements, these methods still rely heavily on extensive unlabeled data to learn prior medical knowledge, e.g., morphology of abdominal CT image.

3   Methodology

3.1   Problem Definition

Within the standard paradigm employed for few-shot segmentation, the dataset is partitioned into two distinct components: the training dataset D_{train} and the testing dataset D_{test}, with their corresponding sets of categories denoted as C_{train} and C_{test} respectively. The class sets of the training and testing datasets are mutually exclusive, hence $C_{train} \cup C_{test} = \emptyset$. Each dataset is structured into episodes, symbolized as $[S, Q]$, where S indicates the support set, and Q refers to the query set. The support set S is formulated as $S = \{(I_s^i, M_s^i(c)), c \in C\}_{i=1}^{K}$, where $I_s^i \in \mathbb{R}^{H \times W}$ is the i_{th} support image, $M_s^i(c) \in \mathbb{R}^{H \times W}$ stands for the ground truth mask for class c of I_s^i, and C is the corresponding class set for the episode $[S, Q]$. Given $|C| = N$, one episode $[S, Q]$ can be defined as a $\boldsymbol{N}$-way-$\boldsymbol{K}$-shot task. The query set Q is analogous to the support set, and is denoted as $Q = \{(I_q, M_q(c)), c \in C\}$.

In the training phase of the few-shot segmentation task, numerous episodes are required within each training epoch. Moving into the testing phase, for each episode $[S, Q] \in D_{test}$, the support set S is utilized to tackle the novel segmentation task, while the query set Q is simultaneously used for evaluation.

Considering the specific challenges in the medical image domain, including a narrow range of tasks and the laborious, costly process of manual image annotation, the conventional few-shot setup falls short of practicality. To address this, we propose a novel few-shot setup tailored to the medical imaging tasks. In medical image segmentation, our goal is to segment target classes labeled as $l \in C_{test} = \{l_1, l_2, \ldots, l_N\}$. For example, if the model currently lacks the capability to segment kidneys, we aim to quickly adapt to kidney segmentation using just a few images and corresponding ground truth masks. To achieve this, we employ a one-way-K-shot support set $S = \{(I_s^i, M_s^i(l))\}_{i=1}^{K}$, which assists in identifying the class l. We set $K = 1$ in our descriptions and experiments, indicating that we use only one medical image and its corresponding segmentation mask as the support set. The query set $Q = \{(I_q^i, M_q^i(l))\}_{i=1}^{|D_{test}|}$ comprises the entirety of the testing dataset D_{test}, excluding the images from the support set S. Notably, our paradigm eliminates the requirement for a training dataset D_{train}, as the one-way-K-shot support set alone imparts the requisite prior information, negating the need for repeated episodes to learn identical classes.

3.2   Revisiting of Segment Anything Model

The Segment Anything Model, or SAM, is designed for promptable segmentation tasks. The task necessitates the generation of a valid segmentation mask based on an image in conjunction with a prompt or a combination of prompts. The prompts can manifest as foreground or background points, bounding boxes, dense masks or free-form text. SAM comprises three components: an image encoder $F(\cdot)$, a prompt encoder $E(\cdot)$, and a mask decoder $D(\cdot)$. The image encoder takes the image as input to create a corresponding feature embedding.

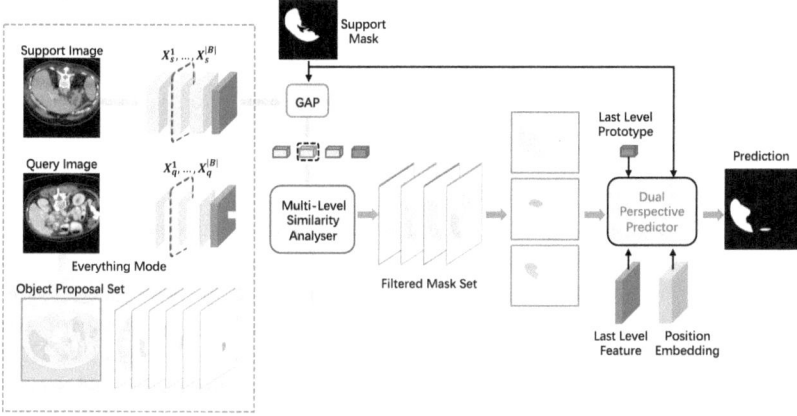

Fig. 1. Overview of our proposed SnapSeg framework.

Concurrently, the prompt encoder creates the prompt embedding from the provided prompts. The mask decoder then uses both of these embeddings to generate the final target mask. In the meantime, SAM provides the "everything mode", which segments the entire image automatically to produce a comprehensive set of masks.

To harness SAM's exceptional zero-shot generalization capability, our method is based on two key assumptions:

– The image encoder $F(\cdot)$ generates generalized features enriched with semantic information.
– SAM is capable of segmenting every required object from any given image, ensuring comprehensive object identification and analysis.

3.3   Overall Framework

Our proposed framework is depicted in Fig. 1. More specifically, the architecture we introduce consists of three distinct modules: (1) generation module: we create a set of masks for a query image as the object proposal set, and create multi-level features for both the support and query images. This process is designed to enrich the information available for subsequent analysis. (2) Multi-level similarity analyser module: we generate similarity maps to measure the relationship between the query image and the support image, and utilize the similarity maps to filter the high-related candidates from the object proposal set. (3) Dual perspective prediction module: The module analyzes correlations between the target class in the support image and the highly related candidates in the query image from both feature and spatial perspectives, and finalizes the mask prediction for the query image.

3.4   Object Proposal Set and Multi-level Features Generator

In our research, we evaluate how various feature levels in the image encoder assess the relationship between the target objects in support and query images. Our results indicate that mid-level features predominantly focus on local positioning. Meanwhile, as feature levels increase, high-level features progressively focus on global distribution and morphology. This shift leads to a stronger ability to define object boundaries clearly and enhance the differentiation between various objects, and results in a diminished emphasis on local positional accuracy.

In medical segmentation, particularly in abdominal organ segmentation, organ sizes in a patient's abdomen increase and then decrease from top to bottom slices. This pattern is especially pronounced in larger organs like the liver and spleen, which show large variability in size and shape across datasets. Conversely, smaller organs such as the kidneys display more uniformity in size and morphology. Thus, organs with substantial morphological variations, like the liver and spleen, benefit from mid-level features that provide localized positional information for accurate segmentation. On the other hand, smaller, more morphologically consistent organs such as the kidneys rely on the distinctiveness offered by high-level features for effective differentiation. It motivates us to propose a framework leveraging multi-level features, ensuring robust and adaptable segmentation performance in diverse medical tasks. Therefore, within the generation module, we leverage the image encoder to produce multi-level features. For clarification, given an episode

$$[S(l) = \{(I_s, M_s(l))\}, Q(l) = \{(I_q, M_q(l))\}]. \tag{1}$$

As the image encoder $F(\cdot)$ in SAM is built on vision transformer, we leverage the outputs from various blocks of the image encoder to derive multi-level features for both the support and query images. We select $|B|$ blocks uniformly from the image encoder's block set to obtain a sampled block set B. It can be formulated as:

$$B = \{b_1, \dots, b_{|B|}\}, \tag{2}$$

where each block yields an output feature represented by $X^k \in \mathbb{R}^{c \times h \times w}$, in which $k \in [1, |B|]$ denotes the index of the chosen block, c denotes the channel dimension, h and w denotes the dimensions of the height and width respectively. Therefore, we can generate two sets of multi-level features for both the support and query images accordingly:

$$[X_s^1, \dots, X_s^{|B|}], \tag{3}$$

$$[X_q^1, \dots, X_q^{|B|}]. \tag{4}$$

In the meantime, we use the "everything" mode of SAM to generate the object proposal set for the query image:

$$\mathcal{P} = \{M_1, M_2, \dots, M_{|\mathcal{P}|}\}. \tag{5}$$

3.5   Multi-level Similarity Analyser

As shown in Fig. 2, to identify the target mask within the object proposal set $\mathcal{P}$, we process the query features in tandem with the support prototype p_s in each level $k \in [1, |B|]$, generating multi-level similarity maps. These similarity maps indicate the likelihood of each pixel in the query image being part of the target class. p_s^k denotes the support prototype for the k_{th} level. Thus, in the k_{th} level, we have:

$$p_s^k = GAP(X_s^k \odot M_s) \in \mathbb{R}^{1 \times c} \tag{6}$$

GAP denotes the global average pooling operation, and $\odot$ is the Hadamard product [5].

For each level in B, we utilize the support prototype p_s as the feature representation of the target class, then we compute the similarity map for the query image to quantify the correlation between the target class and the pixels in the query image,

$$\sigma(i,j)^k = \cos(p_s^k, X_q^k(i,j)), \tag{7}$$

where $i \in [1, h], j \in [1, w]$ and $\sigma^k \in \mathbb{R}^{h \times w}$. σ^k is the similarity map in the k_{th} level. Thus, we could get the multi-level similarity map set:

$$\Sigma = \{\sigma^1, \ldots, \sigma^{|B|}\}.$$

To segment the target class, we take the similarity map set Σ and the object proposal set $\mathcal{P}$ as inputs, and output a filtered mask set, which are the highly correlated candidates for the target class. Specifically, we set a threshold γ for the similarity maps and convert them to binary masks:

$$m^k = t_\gamma(\sigma^k), \tag{8}$$

$$t_\gamma = \begin{cases} 1, & \text{if } x \geq \gamma; \\ 0, & \text{if } x < \gamma. \end{cases} \tag{9}$$

where t_γ is Heaviside step function, it transforms the similarity map into binary mask upon the threshold parameter γ.

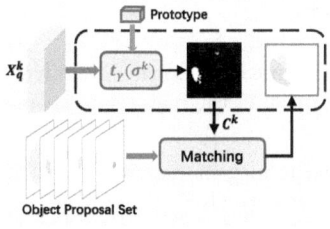

Fig. 2. Similarity Analyser Module

We introduce a matching module that employs morphological analysis to sift through the object proposal set $\mathcal{P}$ and select the most suitable masks based on the binary mask m at each level. In each level, we extract every connected component in the binary mask m: for the k_{th} binary mask m^k, its connected component set of m^k would be

$$C^k = \{c_1^k, \ldots, c_{|C^k|}^k\},$$

where $m^k = c_1^k \cup \cdots \cup c_{|C^k|}^k$. For each connected component c drawn from C^k, we identify at most one mask M from $\mathcal{P}$ that can fully enclose c. This process generates a subset $\mathcal{F}^k = \{M_1, \ldots, M_{|\mathcal{F}^k|}\}$ of $\mathcal{P}$ corresponding to C^k. We compile these subsets for each level in B, resulting in $\{\mathcal{F}^1, \mathcal{F}^2, \ldots, \mathcal{F}^{|B|}\}$. From these, we select the three masks $\{\mathcal{M}_1, \mathcal{M}_2, \mathcal{M}_3\}$ with the highest occurrence counts across all levels as the primary candidates for the subsequent process.

3.6   Dual Perspective Predictor

Aiming for high-confidence mask predictions, we conduct coarse-level analysis at the feature and spatial perspectives. The architecture of dual perspective predictor is illustrated in the Fig. 3. In the perspective of spatial, medical images of a similar category generally exhibit consistent morphological structures. However, due to the medical images are derived from various standardized equipment, discrepancies arise across different images concerning the positioning, alignment, and orientation of the respective organs. To capitalize on this inherent consistency while addressing the aforementioned challenges, we introduce the relative spatial metric method. This approach aims to accurately measure the spatial relationships between highly correlated masks $\{\mathcal{M}_1, \mathcal{M}_2, \mathcal{M}_3\}$ and the ground truth mask $M_s(l)$ for the support image.

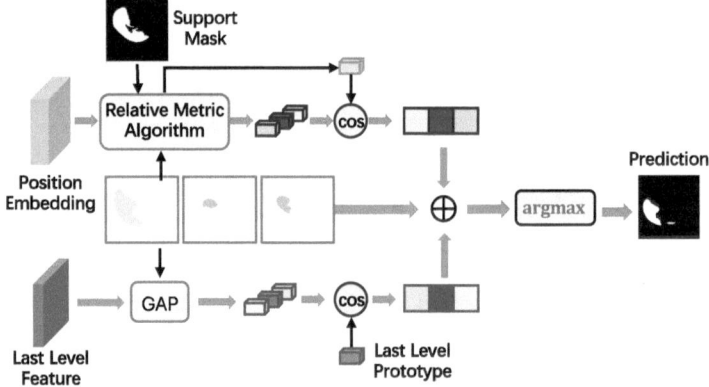

Fig. 3. Dual Perspective Predictor: in the coarse grained level, utilize relative feature and spatial prototypes to find the most matching mask in the high related candidates.

We define an anchor point as the center of the largest component within the medical image, utilizing this anchor point to generate the relative spatial prototype for spatial similarity metrics. Specifically, in the analysis of an abdominal CT image, we employ the abdominal cavity as the anchor mask M_{anchor}, with its mean position embedding acting as the spatial anchor α. The spatial anchor α is determined as follows:

$$\alpha = GAP(\Xi \odot M_{anchor}), \tag{10}$$

where $\alpha \in \mathbb{R}^c$ represents the spatial prototype of the anchor point, and $\Xi \in \mathbb{R}^{c \times H \times W}$ represents the position embedding. To derive the relative spatial prototype for both highly correlated masks $\{\mathcal{M}_1, \mathcal{M}_2, \mathcal{M}_3\}$ and the ground truth mask $M_s(l)$ for the support image, we initially compute their abstract spatial prototypes as follows:

$$\rho_{q,i} = GAP(\Xi \odot \mathcal{M}_i), \mathcal{M}_i \in \{\mathcal{M}_1, \mathcal{M}_2, \mathcal{M}_3\} \tag{11}$$

$$\rho_s = GAP(\Xi \odot M_s). \tag{12}$$

In these equations, $\rho_{q,i}$ defines the spatial prototype for mask $\mathcal{M}_i$ of the query image, while ρ_s identifies the spatial prototype of the support image. By computing the difference from their spatial anchors respectively, we establish the relative spatial prototypes for each mask in both query and support images:

$$\tilde{\rho}_{q,i} = \rho_{q,i} - \alpha_q, \tag{13}$$

$$\tilde{\rho}_s = \rho_s - \alpha_s. \tag{14}$$

This process yields the relative spatial prototypes, sharpening the precision of relative spatial measurements within our analysis. Then we compute the cosine similarity between each $\tilde{\rho}_{q,i}$ and $\tilde{\rho}_s$ to metric the spatial relationship between support target mask and the highly correlated masks $\{\mathcal{M}_1, \mathcal{M}_2, \mathcal{M}_3\}$:

$$\Upsilon_{spat} = [\cos(\tilde{\rho}_{q,1}, \tilde{\rho}_s), \dots, \cos(\tilde{\rho}_{q,3}, \tilde{\rho}_s)]. \tag{15}$$

In the perspective of feature, query prototypes are generated for $\{\mathcal{M}_1, \mathcal{M}_2, \mathcal{M}_3\}$ on the last level $|B|$:

$$p_{q,i} = GAP(X_q^{|B|} \odot \mathcal{M}_i). \tag{16}$$

Then, the similarity metric for the object proposal set at the coarse-grained level is shown as:

$$\Upsilon_{feat} = [\cos(p_{q,1}, p_s^{|B|}), \dots, \cos(p_{q,3}, p_s^{|B|})]. \tag{17}$$

Then, we can predict the target mask of query image based on these two similarity sets:

$$\hat{M}_q \triangleq M[argmax(\Upsilon[1], \Upsilon[2], \Upsilon[3])], \tag{18}$$

$$\Upsilon = \Upsilon_{feat} + \Upsilon_{spat}. \tag{19}$$

4   Experiments

4.1   Dataset and Pre-processing

To evaluate our proposed method, we performed evaluations on the abdominal CT dataset (Abd-CT) [8]. There are 30 Abd-CT Scans in the Abd-CT dataset, each was randomly selected from a combination of one colorectal cancer chemotherapy trial and a retrospective ventral hernia study. The 30 scans have variable volume sizes with resolution varies from $0.54 \times 0.54 \ mm^2$ to 0.98×0.98 mm^2 and variations in intensity distributions between scans. Therefore, the Abd-CT dataset is imhomogeneous, this circumstance allows us to illustrate the general ability of our method.

In our experiment settings, Each scan has undergone standard preprocessing steps, including intensity normalization and resampling. On top of this, each slice in one scan has been transformed into a 2D axial format and resized to 300×300. Meanwhile, In accordance with the input specifications of SAM, each image was replicated three times along the channel dimension.

Table 1. Experimental results (in Dice Coefficient) on Abd-CT images.

method	Spleen	R-Kidney	L-Kidney	Liver	Mean
ALPNet [15]	27.73	30.34	34.96	47.37	35.11
CHNet [23]	63.26	55.41	66.57	73.22	64.62
SSL-RPNet [20]	65.14	66.73	64.01	72.99	67.22
SSL-ALPNet [15]	68.39	66.04	62.14	73.90	67.62
CRAPNet [1]	70.17	67.33	70.91	70.45	69.72
RPT [29]	70.80	67.73	72.99	**75.24**	71.69
Ours	**73.50**	**69.28**	**73.03**	73.93	**72.43**

4.2   Evaluation

For 3D volumetric data in few-shot setting, we follow the Volumetric Segmentation Strategy provided by [16]. Given a designated class, identify the top and bottom slices where the organ of interest appears in the support scan. Subsequently, this region is divided into k uniform parts, each termed as a 'chunk'. The central slice from each chunk is then selected as the support image for that specific position. Given this, every query scan (noting that the query scan and support scan originate from distinct patients) is also uniformly divided into k chunks. For each slice in a chunk, we select the support image from the matched position in its support chunk to establish a $[S, Q]$ pair. We utilize the Dice Coefficient to measure the similarity between the predicted mask and the ground truth of query image. A Dice Coefficient of 100 indicates perfect overlap, while 0 denotes no overlap. Moreover, in our experiments, to maintain consistency with the evaluation configuration of [15], we define four organs as 4 semantic classes: Spleen, Right Kidney, Left Kidney and Liver. All our experiments are performed under one-way one-shot setting.

4.3   Implementation Details

Our method is implemented with PyTorch based on official SAM implementation. We employ the ViT-H model as our base model to obtain a more refined object proposal set. The image encoder is constituted by 32 ViT blocks, we extracted features in $1280 \times 64 \times 64$ size from ViT blocks located at positions $[10, 15, 20, 25, 31]$ to serve as our multi-level features. Each segmentation task takes only 1 s on a single Nvidia GeForce RTX 3090 GPU.

4.4   Quantitative and Qualitative Results

Table 1 showcases a comparative analysis of our SnapSeg framework with state-of-the-art few-shot segmentation methods, including SSL-ALPNet [15], CHNet [23], SSL-RPNet [20], CRAPNet [1], and RPT [29], on the Abd-CT dataset. Distinct from existing methods that predominantly use large volumes of unlabeled data during training to learn prior medical knowledge, SnapSeg directly approaches new segmentation tasks using one labeled image, bypassing the traditional training phase. With SnapSeg achieving a 72.43% Dice Coefficient on Abd-CT and outperforming notable methods such as RPT, CRAPNet and SSL-ALPNet by 0.74%, 2.71%, and 4.81% respectively. This achievement not only establishes a new state-of-the-art but also highlights SnapSeg's potent real-time processing and generalizability. To acquire a more thorough understanding, we further illustrate the visual segmentation results in Fig. 4. The illustration reveals that SnapSeg consistently produces appropriate segmentation of query images by leveraging the corresponding support set.

Table 2. Ablation study results (in Dice Coefficient) on Abd-CT images.

Feat.	Spat.	Spleen	R-Kidney	L-Kidney	Liver	Mean
✓	×	59.95	57.07	60.00	55.70	58.18
✓	✓	73.50	69.28	73.03	73.93	72.43

4.5   Ablation Study

To investigate the impact of spatial branch, especially the relative metric algorithm in our method, we further conduct experiments on the Abd-CT dataset. Adhering to the same implementation setting as mentioned before, we predict the target mask without the spatial branch. As demonstrated in Table 2, the utilization of spatial branch significantly enhances the analysis during coarse-grained comparisons, facilitating a more effective validation of high related candidates.

Moreover, our research extends to evaluating the impact of threshold γ and the number of chunks within the volumetric segmentation strategy on the performance of SnapSeg. Notably, the state-of-the-art (SOTA) methods discussed previously utilize three chunks, a configuration we have maintained in our experiments. As depicted in the two graphs on the right of Fig. 4, setting the threshold

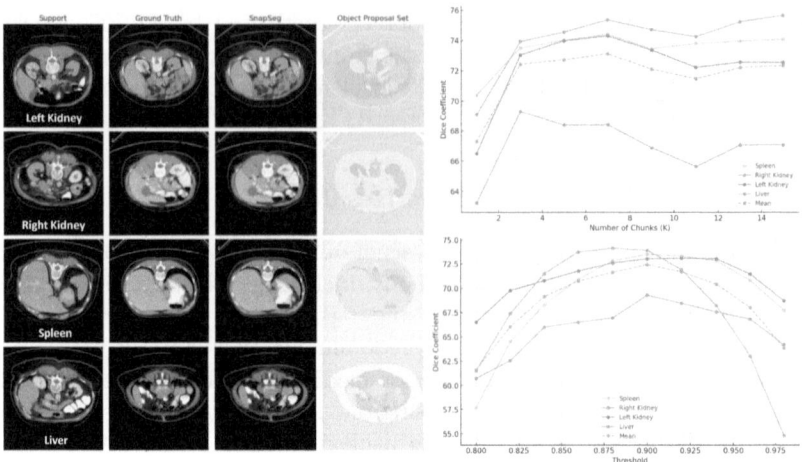

Fig. 4. Left: A qualitative comparison between the segmentation results of our proposed method and the ground truth. Right: The Dice Coefficient vs. the filtering threshold and the number of chunks.

to 0.90 facilitates consistent and balanced segmentation outcomes. Additionally, configuring the volumetric segmentation strategy to incorporate seven chunks enables SnapSeg to reach its optimum Dice Coefficient of 73.11%.

5   Conclusion

We introduce a novel few-shot medical image segmentation method that is effective with just one annotated image. Recognizing the challenges posed by limited labeled data in medical tasks, our approach efficiently utilizes a single example to generalize and segment previously unseen medical images. By incorporating the foundation model SAM, enhanced with multi-level feature fusion, our model captures intricate patterns and structures inherent in medical images. The addition of the Multi-level Similarity Analyzer and Dual Perspective Predictor further boosts the performance of our pre-trained network across various medical segmentation challenges. Looking ahead, we plan to expand this few-shot segmentation paradigm to a broader range of medical conditions, aiming to demonstrate its potential and reliability in practical medical scenarios.

Acknowledgment. This study was supported by the National Key Research and Development Program of China (2023YFC2415400); the National Natural Science Foundation of China (62071210); the Shenzhen Science and Technology Program ($RCYX$20210609103056042); the Shenzhen Science and Technology Innovation Committee ($KCXF Z$2020122117340001); the Guangdong Basic and Applied Basic Research (2021A1515220131); high level of special funds (G030230001) from Southern University of Science and Technology, Shenzhen, China.

References

1. Ding, H., Sun, C., Tang, H., Cai, D., Yan, Y.: Few-shot medical image segmentation with cycle-resemblance attention. In: Proceedings of the IEEE/CVF Winter Conference on Applications of Computer Vision, pp. 2488–2497 (2023)
2. Finn, C., Abbeel, P., Levine, S.: Model-agnostic meta-learning for fast adaptation of deep networks. In: International Conference on Machine Learning, pp. 1126–1135. PMLR (2017)
3. Hansen, S., Gautam, S., Jenssen, R., Kampffmeyer, M.: Anomaly detection-inspired few-shot medical image segmentation through self-supervision with super-voxels. Med. Image Anal. **78**, 102385 (2022)
4. Hao, F., He, F., Cheng, J., Wang, L., Cao, J., Tao, D.: Collect and select: semantic alignment metric learning for few-shot learning. In: Proceedings of the IEEE/CVF international Conference on Computer Vision, pp. 8460–8469 (2019)
5. Horn, R.A.: The hadamard product. In: Proceedings of the Symposium Application Mathematics, vol. 40, pp. 87–169 (1990)
6. Jin, C., Guo, Z., Lin, Y., Luo, L., Chen, H.: Label-efficient deep learning in medical image analysis: challenges and future directions. arXiv preprint arXiv:2303.12484 (2023)
7. Kirillov, A., et al.: Segment anything. arXiv preprint arXiv:2304.02643 (2023)
8. Landman, B., Xu, Z., Igelsias, J., Styner, M., Langerak, T., Klein, A.: MICCAI multi-atlas labeling beyond the cranial vault–workshop and challenge. In: Proceedings of MICCAI Multi-Atlas Labeling Beyond Cranial Vault-Workshop Challenge, vol. 5, p. 12 (2015)
9. Lee, K., Maji, S., Ravichandran, A., Soatto, S.: Meta-learning with differentiable convex optimization. In: Proceedings of the IEEE/CVF Conference on Computer Vision and Pattern Recognition, pp. 10657–10665 (2019)
10. Li, K., Zhang, Y., Li, K., Fu, Y.: Adversarial feature hallucination networks for few-shot learning. In: Proceedings of the IEEE/CVF Conference on Computer Vision and Pattern Recognition, pp. 13470–13479 (2020)
11. Liu, B., et al.: Negative margin matters: understanding margin in few-shot classification. In: Vedaldi, A., Bischof, H., Brox, T., Frahm, J.-M. (eds.) ECCV 2020. LNCS, vol. 12349, pp. 438–455. Springer, Cham (2020). https://doi.org/10.1007/978-3-030-58548-8_26
12. Ma, J., He, Y., Li, F., Han, L., You, C., Wang, B.: Segment anything in medical images. Nat. Commun. **15**(1), 654 (2024)
13. Mangla, P., et al.: Charting the right manifold: manifold mixup for few-shot learning. In: Proceedings of the IEEE/CVF Winter Conference on Applications of Computer Vision, pp. 2218–2227 (2020)
14. Oreshkin, B., Rodríguez López, P., Lacoste, A.: Tadam: task dependent adaptive metric for improved few-shot learning. In: Advances in Neural Information Processing Systems, vol. 31 (2018)
15. Ouyang, C., Biffi, C., Chen, C., Kart, T., Qiu, H., Rueckert, D.: Self-supervision with superpixels: training few-shot medical image segmentation without annotation. In: Vedaldi, A., Bischof, H., Brox, T., Frahm, J.-M. (eds.) ECCV 2020, Part XXIX. LNCS, vol. 12374, pp. 762–780. Springer, Cham (2020). https://doi.org/10.1007/978-3-030-58526-6_45
16. Roy, A.G., Siddiqui, S., Pölsterl, S., Navab, N., Wachinger, C.: 'squeeze & excite'guided few-shot segmentation of volumetric images. Med. Image Anal. **59**, 101587 (2020)

17. Shaban, A., Bansal, S., Liu, Z., Essa, I., Boots, B.: One-shot learning for semantic segmentation. arXiv preprint arXiv:1709.03410 (2017)
18. Snell, J., Swersky, K., Zemel, R.: Prototypical networks for few-shot learning. In: Advances in Neural Information Processing Systems, vol. 30 (2017)
19. Sun, Q., Liu, Y., Chua, T.S., Schiele, B.: Meta-transfer learning for few-shot learning. In: Proceedings of the IEEE/CVF Conference on Computer Vision and Pattern Recognition, pp. 403–412 (2019)
20. Tang, H., Liu, X., Sun, S., Yan, X., Xie, X.: Recurrent mask refinement for few-shot medical image segmentation. In: Proceedings of the IEEE/CVF International Conference on Computer Vision, pp. 3918–3928 (2021)
21. Tian, Z., Zhao, H., Shu, M., Yang, Z., Li, R., Jia, J.: Prior guided feature enrichment network for few-shot segmentation. IEEE Trans. Pattern Anal. Mach. Intell. **44**(2), 1050–1065 (2020)
22. Vinyals, O., Blundell, C., Lillicrap, T., Wierstra, D., et al.: Matching networks for one shot learning. In: Advances in Neural Information Processing Systems, vol. 29 (2016)
23. Wang, B., Li, Q., You, Z.: Self-supervised learning based transformer and convolution hybrid network for one-shot organ segmentation. Neurocomputing **527**, 1–12 (2023)
24. Wang, K., Liew, J.H., Zou, Y., Zhou, D., Feng, J.: PANet: few-shot image semantic segmentation with prototype alignment. In: proceedings of the IEEE/CVF International Conference on Computer Vision, pp. 9197–9206 (2019)
25. Yu, Q., Dang, K., Tajbakhsh, N., Terzopoulos, D., Ding, X.: A location-sensitive local prototype network for few-shot medical image segmentation. In: 2021 IEEE 18th international symposium on biomedical imaging (ISBI), pp. 262–266. IEEE (2021)
26. Zhang, C., Lin, G., Liu, F., Yao, R., Shen, C.: CANet: class-agnostic segmentation networks with iterative refinement and attentive few-shot learning. In: Proceedings of the IEEE/CVF Conference on Computer Vision and Pattern Recognition, pp. 5217–5226 (2019)
27. Zhang, T., Huang, W.: Kernel relative-prototype spectral filtering for few-shot learning. In: Avidan, S., Brostow, G., Cissé, M., Farinella, G.M., Hassner, T. (eds.) ECCV 2022. LNCS, vol. 13680, pp. 541–557. Springer, Cham (2022). https://doi.org/10.1007/978-3-031-20044-1_31
28. Zhmoginov, A., Sandler, M., Vladymyrov, M.: Hypertransformer: model generation for supervised and semi-supervised few-shot learning. In: International Conference on Machine Learning, pp. 27075–27098. PMLR (2022)
29. Zhu, Y., Wang, S., Xin, T., Zhang, H.: Few-shot medical image segmentation via a region-enhanced prototypical transformer. In: Greenspan, H., et al. (eds.) MICCAI 2023. LNCS, vol. 14223, pp. 271–280. Springer (2023). https://doi.org/10.1007/978-3-031-43901-8_26
30. Zhu, Y., Wang, S., Xin, T., Zhang, Z., Zhang, H.: Partition-a-medical-image: Extracting multiple representative sub-regions for few-shot medical image segmentation. arXiv preprint arXiv:2309.11172 (2023)
31. Zhu, Z., He, X., Qi, G., Li, Y., Cong, B., Liu, Y.: Brain tumor segmentation based on the fusion of deep semantics and edge information in multimodal MRI. Inf. Fusion **91**, 376–387 (2023)

Assessing the Generalizability of Cancer Prognosis Models: Breast and Colon Cancer Case Studies

Wafaa Tizi[✉] and Abdelaziz Berrado

Equipe AMIPS - Ecole Mohammadia d'Ingénieurs Mohammed V University in
Rabat, Avenue Ibn Sina, BP765, Agdal, Rabat, Morocco
wafaa.tizi@research.emi.ac.ma

Abstract. Overview Machine learning provides tools to aid in decision-making in various domains. The benefit of these Machine learning techniques is also prevalent in the healthcare industry, especially in cancer prognosis. This paper aims to assess the generalizability of a set of models obtained from a variety of machine learning techniques for survival analysis, by testing the resulting prognosis models on different validation sets of unrelated populations.

Methods Cox Proportional Hazards, Random Survival Forests, and Gradient Boosting for Survival Analysis were used to create breast and colon cancer prognosis models using SEER and SILU datasets respectively. The resulting models were tested on different datasets, namely: Duke and METABRIC breast cancer datasets, and TCGA and CPTAC colon cancer datasets. In order to perform this assessment, we start with a manual preprocessing step of feature matching that resulted in different subsets of the original datasets with a matching feature space. Models were created using each subset of SEER and SILU datasets, and their resulting C-index and Brier score were compared to the subsets of the external validation sets of Duke and METABRIC for breast cancer and TCGA and CPTAC for colon cancer.

Results Different performance metrics give different results. Regardless of the technique used, the performance of the resulting model is not consistent as the validation set changes. The performance of the different models created with the SEER dataset dropped from an average C-index of 70% to 50% on the METABRIC validation set. The choice of the machine learning algorithm for survival analysis can affect the consistency of the resulting models

Keywords: cancer prognosis · survival analysis · machine learning · model validation · generalizability

1 Introduction

Cancer prognosis has been the focus of various research throughout the years, as the main focus of this domain is to predict the survivability of cancer patients

H. Chen et al. (Eds.): TAI4H 2024, LNCS 14812, pp. 123–133, 2024.
https://doi.org/10.1007/978-3-031-67751-9_10

to plan their treatments and follow-up sessions accordingly. Looking at surveys [1,2] conducted on patients with early-stage Breast cancer, 91% of patients wanted to know the results of their prognosis before starting their treatment, while 87% wanted to get their 10-year survival rates [1]. The longest survival time was requested by 85% of the patients, alongside 80% who asked for their 5-year survival rates [2]. These numbers showcase the large interest and importance of tools for cancer prognosis, to satisfy this need, practitioners often rely on their knowledge and/ or the readily available prognosis tools. The abundance of machine learning techniques for cancer prognosis showcases the large interest in developing artificial intelligence models for aided decision-making in this field, in Scopus for example, the number of yearly publications addressing this subject tripled from 20 publications per year in 2013–2014 to over 900 articles in 2023 alone. Regardless of this fact, many papers have questioned the quality of these models and reported on their pitfalls. A systematic review [4] for example, found that the majority of breast cancer prognosis models were less accurate in independent populations as opposed to the development cohort. Another evaluation [3] studied the gastric cancer prediction models to find that insufficient sampling, variable selections, and lack of reporting are a few of the many problems causing the bias of these models. The goal of this paper is to assess the generalizability of cancer survival analysis algorithms and models by studying the potential fluctuation in their performance as the validation set changes. The rest of this paper is organized as follows, Sect. 2 summarizes a collection of studies presenting the pitfalls and problems regarding the generalizability of machine learning models. Section 3 presents the experimental setup, the datasets used, and the

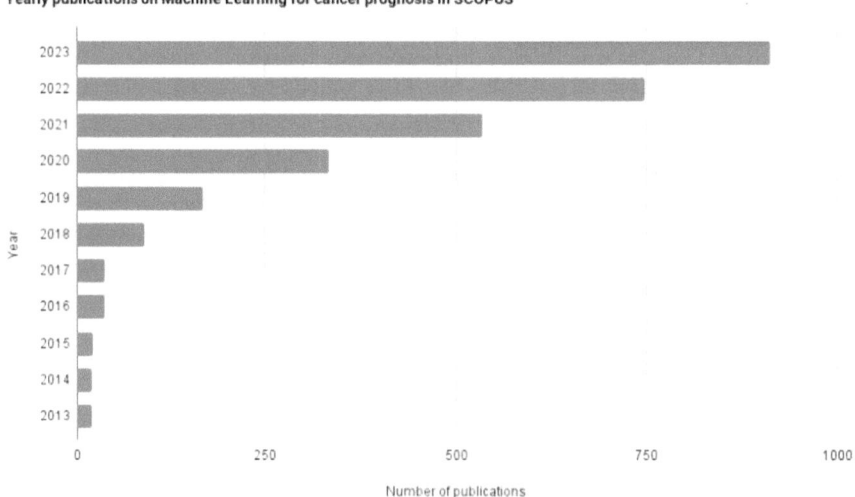

Fig. 1. Trends in the yearly publications on Machine learning for cancer prognosis in SCOPUS

process of feature matching used. Section 4 showcases the results and analyzes the noticeable trends. Finally, Sect. 5 is for conclusions and recommendations (Fig. 1).

2   Literature Review on the Generalizability of Machine Learning Models

Due to the scarcity of articles addressing the generalizability of machine learning models for the case of cancer prognosis, we searched for articles that reported on machine learning models' performances, potential limitations, and/or pitfalls. We conducted a search in the Scopus database for keywords such as: "machine learning", "model validation", and "generalizability". Within the references of the resulting papers, we manually searched for potential studies relevant to this topic. Maleki et al. [5] Conducted a study that evaluates problems within the methods adopted in creating machine learning models that hinder their widespread adoption in clinical practice: the negligence of the independence assumption, problems related to the reporting and evaluation using inappropriate performance indicators, and batch effect. This study concluded that these methodological pitfalls cannot be well assessed using internal validation techniques that often result in inaccurate predictions and misleading results, hindering the development of generalizable models. Dexter et al. [6] Attributed the lack of generalizability of their machine learning models to the influence of the variant syntactic nature of free text datasets across the various laboratory systems. With a focus on acute kidney injury the study Cao et al. [7] benchmarked the performance of a created model to evaluate its generalizability in different subgroups. The results found by this study highlighted the complexity of discrepancies in the performance of artificial intelligence models in subgroups. Most importantly some of these inconsistencies cannot be directly attributed to nor explained by the sample size of the model training. Azad et al. [8] argued that the poor quality of reporting hinders the ability to replicate, reproduce, and externally validate clinical prediction models. This paper advised closely following the Transparent Reporting of a multivariable prediction model for Individual Prognosis Or Diagnosis (TRIPOD) guidelines and optimizing models using overfitting prevention techniques and external validation studies.

3   Experimental Setup

In this section we present the design of our experiment, the choice of algorithms, the datasets used, and how we utilized a process of feature matching to create the necessary data for our assessment.

3.1   Algorithms

The aim of this study is to evaluate how machine learning prognosis models, created by different algorithms, would generalize in an independent population, for this purpose, we picked a collection of machine learning algorithms for survival analysis, namely: Cox Proportional Hazards, Gradient Boosting for Survival Analysis and Random Survival Forests. These techniques have been widely used in studying the survivability of cancer patients and have been benchmarked in various studies [9, 11].

3.2 Datasets

To verify how well a machine learning model for survival analysis would generalize from one population to another, we need different datasets, with matching feature spaces, that were collected from different populations. Due to the lack of these requirements in public and available cancer survival datasets, we opted for a process of feature matching: From a collection of breast and colon cancer datasets, we create subsets that only contain the matching features. Table 1 presents three breast cancer survival datasets: SEER, Duke, and METABRIC. Each dataset contains a variety of features, some of which are common features between them. Table 2 presents three Colon cancer datasets and the resulting subsets containing the matched features. Since the goal was to create subsets of these datasets with the maximum amount of matching features, for the case of Breast cancer, we opted for the creation of 4 dataset subsets instead of just 3 that include two subsets of SEER dataset with the maximum number of features with the Duke and METABRIC datasets respectively. These changes in the feature space are marked with * in Table 2. This process of finding matching features creates a major loss of information but is necessary for our empirical comparison in order to answer the question of how a survival model's performance changes with the change of the chosen algorithm and when the validation population changes. The SEER dataset is used as the reference for the Breast cancer case study and will be used to create models that will be tested on a test set from SEER itself, and validation sets from Duke and METABRIC. The SiLU dataset is used as the reference for the Colon cancer case study and will be used to create models using the different algorithms that will be tested on test sets from SiLU itself and validation sets from CPTAC and TCGA.

4   Empirical Comparison

In this section, we explain the design of the experiment. First, we start by presenting the data processing step, followed by hyperparameter tuning and the result and analysis of the experiment.

4.1   Data Preprocessing

The SEER and METABRIC datasets required little to no cleaning as they are properly processed benchmarking datasets. When it comes to the Duke breast cancer dataset, MRI data columns were removed since we only need the clinical information for our study. The process of feature matching is straightforward for

Table 1. Matching features between the different Colon cancer datasets

Sidra-LUMC (SiLU) Colon Cancer [16]	CPTAC-2 Colon Cancer dataset [17]	TCGA Colorectal Adenocarcinoma[18]
Country: The Netherlands	Country: United States of America	Country: Canada & United Kingdom
Cancer Type: • Colon Adenocarcinoma	Cancer Type: • Colon Adenocarcinoma	Cancer Type: • Colorectal Adenocarcinoma
Samples: 348	Samples: 110	Samples: 640
Matching features: → Age → N Stage → Cancer Stage → T Stage → Fraction Genome Altered → Sex → Tumor Location → Histology → Overall Survival → Overall Survival Status		

Table 2. Matching features between the different Breast cancer datasets

Seer Breast Cancer [13]	Duke- Breast- Cancer- MRI [14]	METABRIC [15]
Country: United States of America	Country: United States of America	Country: Canada & United Kingdom
Cancer Type: • Invasive ductal carcinoma • Invasive lobular carcinoma	Cancer Typee: • Ductal carcinoma in situ • Invasive ductal carcinoma • Invasive lobular carcinoma • Metaplastic • Lobular carcinoma in situ • Tubular, mixed type • Micropapillary • Colloid	Cancer Type: • Breast Invasive Ductal Carcinoma • Breast Mixed Ductal and Lobular Carcinoma • Breast Invasive Lobular Carcinoma • Breast Invasive Mixed Mucinous Carcinoma • Metaplastic Breast Cancer
Samples: 4024	Samples: 596	Samples: 1904
Matching features: → Age → Race → T stage* → N Stage* → Estrogen Status → Progesterone Status → Regional Node Positive* → Tumor Size* → Survival Months → Status	Matching features: → Age → Race → T stage* → N Stage* → Estrogen Status → Progesterone Status → Survival Months → Status	Matching features: → Age → Race → Estrogen Status → Progesterone Status → Regional Node Positive* → Tumor Size* → Survival Months → Status

common features such as Age, ER, and PR status. Rows with missing data were deleted and we chose not to act on the distribution of the data. Similarly for the Colon cancer datasets, we removed common columns with a large number of missing values and kept only the commonly available features, some of which were extracted from available patients' information. From the TCGA dataset, we extracted patients with Colon Adenocarcinoma and excluded rows containing patients with a different cancer type.

4.2   Hyperparameter Tuning

The Gradient Boosting for Survival analysis, Cox Proportional Hazards, and Random Survival Forests machines were used to train different models using the two different subsets of the SEER breast cancer dataset that were obtained through the process of feature matching addressed in previous sections. To optimize the hyperparameters we used cross-validation with random shuffle splits of 80/20, and grid search on a random pool of hyperparameter values. We optimized the number of estimators, learning rate, and depth of the trees used in Gradient Boosting for Survival analysis. For Random Survival Forests we optimized the number of trees and their depth. The Cox Proportional Hazards method we used includes a Ridge penalty that was optimized, and different methods for dealing with tied events such as the Breslow method [12]. The performance measure used to compare is the mean of Harrell's Concordance Index and Brier Score. All the implementations used are available in the Scikit-Survival library 0.18.0. Tables 3 and 4 present a summary of the work conducted in this study. It contains the results of the Mean Concordance Index and Brier score of each implementation alongside the Standard Deviation (Std).

4.3   Results and Analysis

As explained in the previous sections and illustrated by Table 2, two subsets of the SEER datasets were trained independently using the three machine learning techniques of Random Survival Forests, Gradient Boosting for Survival Analysis, and Cox Proportional Hazards with ridge regression. The 6 resulting models were validated on different external datasets of Duke and METABRIC breast cancer datasets. The validation performance was compared to the mean performance of each model during Cross-validation. The first two columns of Table 4, from the left, represent the comparison between the results obtained with the first 3 models trained and tested on the first subset of the SEER dataset in comparison with the results obtained on the validation dataset of Duke.

The results show that when the concordance index is used to measure the performance, the latter stays consistent between the test and validation sets regardless of the algorithm used.

However, the Brier score captured a slight drop in the performance in the Cox proportional hazards model and the Gradient Boosting for Survival Analysis model when used on the validation set of Duke. The model obtained with the Random Survival Forests algorithm presents a consistent performance on both the SEER and Duke datasets.

Overall for the first test, the performance of the models stayed somewhat consistent when tested on a subset of the SEER training set in comparison to the external validation set of Duke.

The second test, columns 3 and 4, present the results obtained when the second subset of the SEER dataset, with matching features with the METABRIC dataset. The performance in this setting is very different from the previous one, since in this case, we notice a large drop in the performances of all three models

Table 3. Mean C-index and Brier score obtained for the subset of the SiLU dataset and the subset validation sets of CPTAC and TCGA.

			COLON CANCER		
			Training set: SILA Validation set: SILA	Training set: SILA Validation set: TCGA	Training set: SILA Validation set: CPTAC
Gradient Boosting for Survival Analysis	C-index	Mean	0.83	0.75	0.587
		Std	0.08	0.01	0.06
	Brier score	Mean	0.133	0.20	0.09
		Std	0.047	0.0084	0.03
Cox Proportional Hazards	C-index	Mean	0.77	0.76	0.65
		Std	0.09	0.03	0.01
	Brier score	Mean	0.167	0.15	0.09
		Std	0.05	0.01	0.02
Random Survival Forests (RSF)	C-index	Mean	0.81	0.76	0.59
		Std	0.077	0.015	0.05
	Brier score	Mean	0.156	0.187	0.09
		Std	0.03	0.007	0.023

when tested on a subset of the training set and when tested on the validation set of METABRIC, from a 70% concordance on the SEER testing set, to a 50% concordance on the external validation set of METABRIC, which is a value that indicates that the models did no better than random guessing. The Brier score, however, showed that the models' performances stayed consistent when tested on the subset of the training set and the validation set.

Regarding the results obtained from the Colon cancer case study, presented in Table 3, the drop in performance was also noticeable as the population changed especially for models created with the Gradient Boosting for Survival Analysis and Random Survival Forests. These methods, when tested internally, using the C-index, outperformed the model obtained using the Cox Proportional Hazards model, however, the latter managed to produce consistent results in comparison with the two ensemble methods. In the CPTAC dataset, the results obtained by the two ensemble methods displayed a major drop in performance from an 80% C-index on the SiLU dataset to 59%. While the Cox PH model performance dropped from 77% to 65%. Similar to the Breast cancer case study, the results

Table 4. Mean C-index and Brier score obtained for each subset of the SEER dataset and the subsets of the validation sets of DUKE and METABRIC.

			BREAST CANCER				
			Training set: SEER / Validation set: SEER	Training set: SEER / Validation set: Duke		Training set: SEER / Validation set: SEER	Training set: SEER / Validation set: METABRIC
Gradient Boosting for Survival Analysis	C-index	Mean	0.74	0.73		0.70	0.48
		Std	0.04	0.003		0.02	0.005
	Brier score	Mean	0.149	0.171		0.12	0.11
		Std	0.02	0.004		0.007	0.007
Cox Proportional Hazards	C-index	Mean	0.73	0.78		0.71	0.50
		Std	0.05	0.002		0.02	0.002
	Brier score	Mean	0.145	0.187		0.11	0.09
		Std	0.02	0.003		0.006	0.003
Random Survival Forests (RSF)	C-index	Mean	0.722	0.714		0.71	0.5
		Std	0.04	0.002		0.02	0.003
	Brier score	Mean	0.148	0.147		0.11	0.09
		Std	0.02	0.002		0.006	0.003

obtained with the Brier score tend to differ from the ones obtained using the C-index.

These results, clearly showcase the effect of the choice of the performance measure on the results obtained. In the Breast cancer case study, the C-index failed to capture the slight drop in performance that we can notice in the Brier score. Contrarily, the C-index captured a significant drop in performance, from the testing set to the validation set, that the Brier score did not manage to portray.

The Colon cancer case study emphasized the potential impact of the choice of the machine learning algorithm's choice on the consistency of the resulting models. The results displayed in Table 3 demonstrate that, although the Cox PH model is outperformed by the models obtained by the two ensemble methods when tested internally, the latter models failed to generalize as well as the Cox PH model on the independent population of CPTAC.

The other pressing analyses of the results are directly linked to the purpose of our study. When validating the Breast cancer models with the first external dataset of Duke, the results gave a misleading level of confidence that the models' performance was consistent even in the external validation set. However, we soon noticed that with the change in the external validation data used, the models became less performant in comparison.

This noticeable shift in results may be due to the location of data collection between the SEER, Duke, and METABRIC datasets, as this is the only major difference that can potentially explain the drop in the models' performances. A similar remark can be noticed in the Colon cancer datasets as the SiLU and CPTAC datasets were collected on different populations of the United States of America and The Netherlands. However, the TCGA dataset contained a broader spectrum of patient samples.

5   Conclusions

In this study, we compare the performances of different Breast and Colon cancer survival models. From the three datasets of SEER, Duke, and METABRIC Breast cancer datasets, and the three SiLU, CPTAC, and TCGA Colon cancer datasets, we created subsets that contained only the matching features. We then used three popular algorithms, namely, the Cox Proportional hazards technique with ridge regression, Gradient Boosting for survival analysis, and Random Survival Forests. These techniques were used to train models using the subsets of a reference dataset, SEER for Breast cancer and SiLU for Colon cancer. The mean of these models' performances was then compared to the results obtained on the external validation sets of Duke and METABRIC Breast cancer datasets and CPTAC and TCGA Colon cancer datasets. A somewhat consistent performance was noticed between the SEER and Duke datasets, except for a slight drop that was captured only when the Brier score was used as a performance measure. A large drop in performance was noticed between the SEER and METABRIC datasets when compared with the Concordance index. Similarly, the results of the C-index captured a major drop in the performance of models from the SiLU to the CPATC dataset, which was not captured with the Brier score. These results indicate that the choice of performance measure has a big influence on the interpretation of model performances. The Colon cancer case study showed that the choice of the algorithm to train the survival models could have an impact on its generalizability, regardless of how well these models might perform when tested internally. On the other hand, the external validation performance depends on the validation set used. The latter can have a significant impact on the assessment of the generalizability of the models. Our study indicates that the reporting of the performance of survival models may be more accurate if the performance of a variety of external validation sets is reported as well. That being said, caution should be exercised when using pre-trained models that are readily available online, especially when there is a mismatch in the location of the data collection. One potential solution for this problem is to opt for personalized models that are trained on local data or testing external models on a sample of internal retrospective patients' data collection. This paper presents a preliminary result with the aim of suggesting to practitioners and patients that some survival estimation models may not accurately predict survivability, especially if the models are trained on different populations. A further analysis that includes a broader spectrum of cancer survival datasets with different data collection time frames will be conducted to dive deeper into the assessment of the generalizability of the cancer survival models.

Acknowledgments. This work was supported by the Ministry of Higher Education, Scientific Research and Innovation, the Digital Development Agency (DDA) and the CNRST of Morocco (Alkhawarizmi /2020/12).

Disclosure of Interests. The authors declare that they have no known competing financial interests or personal relationships that could have appeared to influence the work reported in this paper.

References

1. Lobb, E.A., Kenny, D.T., Butow, P.N., Tattersall, M.H.: Women's preferences for discussion of prognosis in early breast cancer. Health Expect. **4**(1), 48–57 (2001)
2. Hagerty, R.G., et al.: Cancer patient preferences for communication of prognosis in the metastatic setting. J. Clin. Oncol. **22**(9), 1721–30 (2004)
3. He, S., et al.: Real-world practice of gastric cancer prevention and screening calls for practical prediction models. Clin. Transl. Gastroenterol. (2023). https://doi.org/10.14309/ctg.0000000000000546
4. Phung, M.T., Tin Tin, S., Elwood, J.M.: Prognostic models for breast cancer: a systematic review. BMC Cancer (2019). https://doi.org/10.1186/s12885-019-5442-6
5. Maleki, F., Ovens, K., Gupta, R., Reinhold, C., Spatz, A., Forghani, R.: Generalizability of machine learning models: quantitative evaluation of three methodological pitfalls. Radiol. Artif. Intell. **16;5**(1), e220028 (2022). https://doi.org/10.1148/ryai.220028. PMID 36721408; PMCID PMC9885377
6. Dexter, G.P., Grannis, S.J., Dixon, B.E., Kasthurirathne, S.N.: Generalization of machine learning approaches to identify notifiable conditions from a statewide health information exchange. AMIA Jt Summits Transl. Sci. Proc. **2020**, 152–161 (2020)
7. Cao, J., et al.: Generalizability of an acute kidney injury prediction model across health systems. Nat. Mach. Intell. **4**(12), 1121–1129 (2022)
8. Azad, T.D., et al.: Fostering reproducibility and generalizability in machine learning for clinical prediction modeling in spine surgery. Spine J. (2020). https://doi.org/10.1016/j.spinee.2020.10.006
9. Moncada-Torres, A., van Maaren, M.C., Hendriks, M.P., Siesling, S., Geleijnse, G.: Explainable machine learning can outperform Cox regression predictions and provide insights in breast cancer survival. Sci. Rep. (2021). https://doi.org/10.1038/s41598-021-86327-7
10. Kim, H., Park, T., Jang, J., Lee, S.: Comparison of survival prediction models for pancreatic cancer: cox model versus Machine Learning models. Genom. Inform. (2022). https://doi.org/10.5808/gi.22036
11. Tizi, W., Berrado, A.: Machine learning for survival analysis in cancer research: a comparative study. Sci. Afr. **21**, e01880 (2023). https://doi.org/10.1016/j.sciaf.2023.e01880
12. Breslow, N.: Covariance analysis of censored survival data. Biometrics **30**, 89–99 (1974)

13. Jing, T.: SEER breast cancer Data. IEEE Dataport (2019). https://doi.org/10.21227/a9qy-ph35

14. Saha, A.: A machine learning approach to radiogenomics of breast cancer: a study of 922 subjects and 529 DCE-MRI features. Br. J. Cancer (2018). https://doi.org/10.1038/s41416-018-0185-8

15. Pereira, B., et al.: The somatic mutation profiles of 2,433 breast cancers refines their genomic and transcriptomic landscapes. Nat. Commun. (2016). https://doi.org/10.1038/ncomms11479

16. Roelands, J., et al.: An integrated tumor, immune and microbiome atlas of colon cancer. Nat. Med. **29**(5), 1273–1286 (2023). https://doi.org/10.1038/s41591-023-02324-5

17. Vasaikar, S., et al.: Proteogenomic analysis of human colon cancer reveals new therapeutic opportunities. Cell **177**(4), 1035-1049.e19 (2019). https://doi.org/10.1016/j.cell.2019.03.030

18. cbioportal.org/study/summary?id=coadreadtcga

SISU: A Holistic Self-training Framework on Semi-supervised White Blood Cell Segmentation

Hien Quang Kha[1,2], Minh Huu Nhat Le[1,2], Lam Huu Phuc Nguyen[2], Minh Nguyen Tuan Tran[1,2], Linh My Nguyen[6], Hung Quay Thong[7], and Nguyen Quoc Khanh Le[1,2,3,4,5(✉)]

[1] International Ph.D. Program in Medicine, College of Medicine, Taipei Medical University, Taipei 110, Taiwan
khanhlee@tmu.edu.tw
[2] AIBioMed Research Group, Taipei Medical University, Taipei 110, Taiwan
[3] Professional Master Program in Artificial Intelligence in Medicine, College of Medicine, Taipei Medical University, Taipei 110, Taiwan
[4] Research Center for Artificial Intelligence in Medicine, Taipei Medical University, Taipei 110, Taiwan
[5] Translational Imaging Research Center, Taipei Medical University, Taipei 110, Taiwan
[6] International Master of Business Administration Program, College of Management, National Taipei University of Technology, Taipei 106344, Taiwan
[7] Department of Applied Foreign Languages, National Taipei University of Business, Taipei 100, Taiwan

Abstract. The automated segmentation of leukocytes plays a vital role in diagnosing and monitoring life-threatening conditions such as leukemia and lymphoma. However, this task encounters challenges due to the limited availability and quality of public datasets. To effectively utilize the limited dataset, it is necessary to develop a semi-supervised learning framework. In this regard, we propose a holistic self-training framework called Self-training using In-turn Supervised-Unsupervised training (SISU). Our framework encompasses two key components. Firstly, we introduce Feature Perturbed Cross-View Co-Training, which incorporates dual feature perturbation methods and utilizes two auxiliary decoders to enhance the robustness of the feature representation. Secondly, drawing inspiration from FixMatch, we integrate a regularization mechanism via weak-to-strong consistency training to further enhance the self-training framework. Finally, we conduct training and evaluation of SISU for semi-supervised learning on three datasets (Zheng 1, Zheng 2, and LISC), achieving remarkable mIOU scores of up to 93.54%, 88.92%, and 77.02% respectively across multiple settings within the self-training scheme.

Keywords: Semi-supervised learning · Consistency Regularization · FixMatch · Leukemia · Lymphoma · White blood cell

© The Author(s), under exclusive license to Springer Nature Switzerland AG 2024
H. Chen et al. (Eds.): TAI4H 2024, LNCS 14812, pp. 134–144, 2024.
https://doi.org/10.1007/978-3-031-67751-9_11

1   Introduction

Lymphoma and leukemia are amongst the most common and lethal blood cancers [1] originating in the bone marrow. These diseases cause an overproduction of abnormal white blood cells (WBCs), crucial components of the body's immune system. The rapid accumulation of these defective cells crowds out healthy red blood cells, platelets, and normal white cells, leading to various health issues. These include anemia due to low red blood cell counts, excessive bleeding and bruising from insufficient platelets, and an increased risk of infections due to a lack of functional white cells.

Blood microscopy is vital for diagnosing leukemia and lymphoma. Traditional microscopy methods in laboratories, involving manual counting and classification of WBCs, are time-consuming and require high expertise and precision due to the variability in cell images and staining techniques. This variability poses significant challenges in consistently identifying and classifying different WBC types, making the process less efficient and potentially less accurate [2]. Therefore, developing automated systems that can reliably perform these tasks is essential for improving diagnostic workflows and outcomes in hematological analyses.

Similar to various healthcare problems, accurate segmentation of leukocytes is foundational for effective diagnosis, particularly in automated systems that identify and classify different WBC types, a key step in diagnosing and monitoring these diseases [3,14]. Both manual and automated WBC segmentation techniques face significant challenges, especially under real-world conditions. Manual segmentation demands high expertise and is time-consuming, potentially delaying crucial diagnoses and suffering from inconsistencies due to human variability [3,14]. Automated segmentation faces difficulties due to the insufficient quality and quantity of public datasets, which often fail to represent the diversity encountered in clinical environments. These datasets are typically not robust enough to train algorithms that perform accurately across different settings, further complicated by the technological limits of imaging and staining methods [2,14]. Additionally, automated methods usually require extensive, varied, and annotated data, which are rare in public databases, complicating the development and generalization of these systems across the broad spectrum of real-world scenarios [2,14].

To address the lack of annotated datasets on WBCs for training an automated diagnostic system for leukemia and lymphoma, we employ the semi-supervised learning technique. Even with a limited amount of labeled data, semi-supervised learning has long been demonstrated to effectively utilize unlabeled datasets and be capable of constructing powerful models [11].

This work proposes a holistic self-training framework, called Self-training using In-turn Supervised-Unsupervised training (SISU). Similar to a plain self-training scheme, our framework focuses on a semi-supervised learning setting by leveraging unlabeled data to boost the performance of the white blood cell segmentation. **First**, we designed Feature Perturbed Cross-View Co-Training that performed two methods of feature perturbation and fed through two auxiliary decoders for co-training. **Secondly**, drawing inspiration from FixMatch [18], we incorporated a regularization mechanism using weak-to-strong consistency training to enhance the self-training framework. **Finally**, we trained and evaluated the performance of our self-training approach for semi-supervised learning across various settings on three datasets including Zheng 1 [20], Zheng 2 [20], and LISC [21], achieving state-of-the-art results across multiple settings in the self-training scheme.

2   Related Works

Medical Image Segmentation. Recent advancements in AI-based diagnostic systems have significantly contributed to anomaly detection through medical imaging domains, such as radiology [9,10,12,13], microscopy [15]. Among these systems, segmentation models, particularly U-net-based [16] models, play crucial roles in accurately predicting lesions. The integration of U-net with Resnet-family [17] backbones has further enhanced their performance, making this combination one of the most widely used baselines for addressing semantic segmentation challenges.

Semi-supervised Semantic Segmentation. In the field of supervised image segmentation, a pressing challenge lies in the scarcity of labeled data, primarily due to the labor-intensive process of annotating each pixel within an image. Nonetheless, a complete transition to unsupervised learning poses the risk of sacrificing precision and accuracy, particularly detrimental in medical contexts. In addressing this problem, the emergence of semi-supervised learning, featuring techniques such as consistency regularization [18] and pseudo-labeling [22], proves instrumental in facilitating image segmentation with limited labeled data while leveraging unlabeled data. With semi-supervised semantic segmentation as the backbone, the self-learning-based approach allows the computer to fine-tune the model by using the labeled image as guidance to gradually turn unlabeled images into labeled training data through many iterations. Within this framework, the Self-training ST++ methodology [22] distinguishes itself by utilizing selective re-training based on prediction-level stability rather than indiscriminately incorporating all pseudo-labeled data.

3   Methodology

3.1   Stage 1: Feature Perturbed Cross-View Co-training Scheme

In the first stage, supervised learning is employed to train on labeled dataset D_l. Cross-entropy (CE) based supervised loss L_s is used to train the segmentation network f given a labeled training example x_i^l and its pixel-level label y_i as:

$$L_{s_1} = \frac{1}{|D_l|} \sum_{x_i, y_i \in D_l} CE\left(y_i, f\left(x_i^l\right)\right)$$

The shared encoder $z_i = h(x_i^u)$ is used to compute an intermediate representation of the input for an unlabeled example, x_i^u. In order to feed the two perturbed versions to the two auxiliary decoders, we produce two perturbed versions z_i^1 and z_i^2 of the intermediate representation z_i. We treat the perturbation function as a component of the auxiliary decoder for consistency training. In detail, two perturbed features are generated by using Feature DropOut and Feature Noise-Injection for FP_1 and FP_2, respectively. Next, the training goal is to reduce the unsupervised loss L_u, which quantifies the difference between the output of the auxiliary decoders and the main decoder:

$$L_{u_1} = \frac{1}{2|D_u|} \sum_{x_i \in D_u} \sigma\left(g\left(z_i\right), g_a^1\left(z_i\right)\right) + \sigma\left(g\left(z_i\right), g_a^2\left(z_i\right)\right)$$

whereas g and σ are generators and the mean squared error (MSE) for measuring the distance between two output probability distributions, respectively. The final loss L combination for the first stage of training is defined as:

$$L = L_{s_1} + w_{u_1} L_{u_1}$$

with L_{s_1} as the supervised loss and w_{u_1} as a hyperparameter denoting unsupervised loss of weight in the first stage.

3.2   Stage 2: Weak-to-Strong Consistency Self-training

After the first training stage, the model gained a substantial amount of prior knowledge from both labeled and unlabeled datasets. However, confirmation bias is still a concern. To reduce the effect of noisy pseudo-labeling, we design the strategy of a self-training method similar to prior curriculum learning [19]. Our method aims to strategically select reliable samples for supervised training before doing weak-to-strong consistency training, which is inspired by FixMatch [18].

First, the predicted images from the first stage are used to filter the high-quality prediction using mean IOU (mIOU) estimation. In particular, we find a favorable association between segmentation performance and the developing stability of created pseudo masks throughout the supervised training phase. As a result, the most dependable and accurately predicted unlabeled images can be chosen based on their evolving stability during training. In addition to reflecting the dependability of the pseudo mask and unlabeled picture, the mIOU can be used as a stability metric as below:

$$s_i = \sum_{j=1}^{K-1} mIOU\left(M_{ij}, M_{iK}\right)$$

where s_i is the stability score, reflecting the reliability of u_i, and K is the checkpoint saved during the training stage.

After selecting the partially reliable samples, these pseudo-label masks are used for supervised training again similar to the previous stage as the following formula:

$$L_{s_2} = \frac{1}{|D_{l_2}|} \sum_{x_i, y_i \in D_{l_2}} CE\left(y_i, f\left(x_i^{l_2}\right)\right)$$

where D_{l_2} is the sum of labeled dataset D_l and reliable pseudo-label dataset.

Then, the model is continuously trained under an unsupervised setting using weak-to-strong consistency training. To be specific, for C is the number of classes to predict, the weak-to-strong scheme aims to estimate a pseudo-label $\tilde{y} \in R^C$ for each strongly augmented unlabeled data x in D_u. The unsupervised loss can be defined as:

$$L_{u_2} = \frac{1}{N_l} \frac{1}{|D_u|} \sum_{x \in D_{u_2}} \sum_{i=0}^{N_l - 1} CE\left(\tilde{y}_i, f_\theta\left(S \cdot w\left(x_i\right)\right)\right)$$

whereas N_l is the number of valid labeled pixels in an image x, $|D_{u_2}|$ is the cardinality of the remaining unlabeled dataset, contains unreliable unlabeled dataset. S is the strong augmentation strategy operator. In detail, we used color-jittering and random CutOut operations as strong augmentation operators. $f_\theta\left(w\left(x_i\right)\right)$ is the predicted probability of pixel i.

3.3   Stage 3: All Samples Supervised Re-training

Finally, after performing co-training and consistency unsupervised training schemes, the recently updated model is used to predict the remaining unlabeled data D_{u_2} to generate the pseudo-mask. The whole dataset, now, has full labels including original masks and high-quality pseudo-masks. All data is lastly trained through supervised learning, that computed as:

$$L_{s_3} = \frac{1}{|D_{l_3}|} \sum_{x_i, y_i \in D_{l_3}} CE\left(y_i, f\left(x_i^{l_3}\right)\right)$$

where D_{l_3} indicates all original and high-quality pseudo-label sets.

4   Experiments

4.1   Dataset

Zheng 1 [20] consists of 300 120 × 120 sub-images of individual WBCs (176 neutrophils, 22 eosinophils, 1 basophil, 48 monocytes, and 53 lymphocytes) obtained from 80 source images. The source images were captured using a Motic Moticam Pro 252A optical microscope camera with a N800-D motorized auto-focus microscope. The smears were processed using a newly-developed hematology reagent for rapid WBC staining. Each image in the dataset has a size of 2048 × 1536 and a color depth of 24 bits.

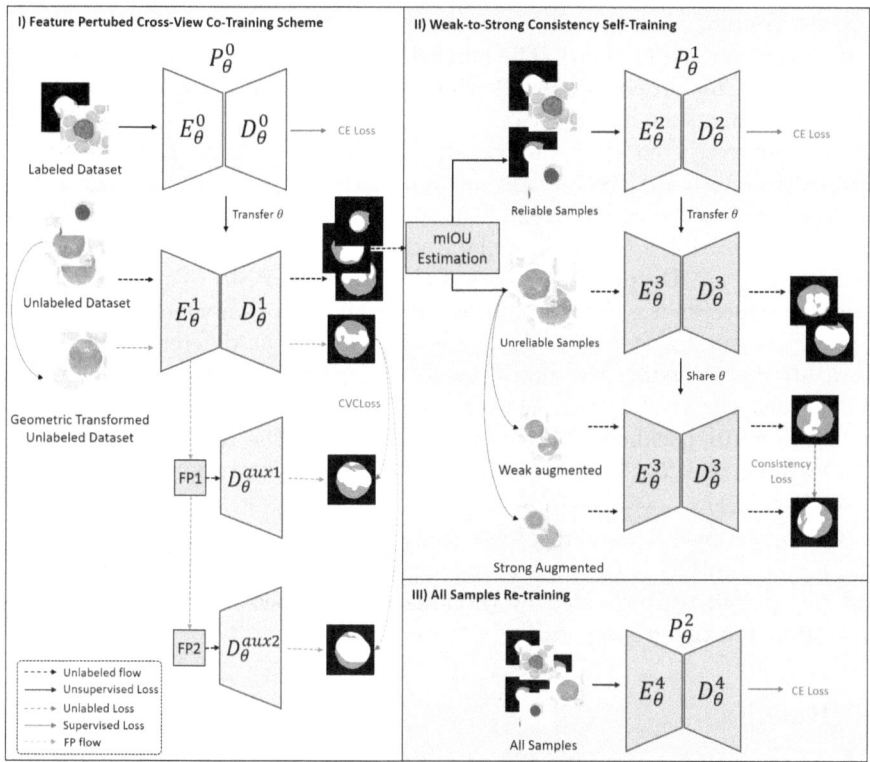

Fig. 1. Our proposed SISU semi-supervised semantic segmentation framework. In the first stage, feature perturbed cross-view co-training is incorporated to boost the capabilities of model on an unlabeled dataset. The second stage focuses on selecting reliable samples for supervised training and improving robustness by using weak-to-strong consistency training. Finally, the model is trained under a fully labeled dataset using original masks and high-quality generated masks.

Zheng 2 [20] comprises 100 color images, each sized 300×300 pixels, obtained from the CellaVision blog. The images consist of 30 neutrophils, 12 eosinophils, 3 basophils, 18 monocytes, and 37 lymphocytes (Fig. 1).

The LISC dataset [21] contains hematological images taken from the peripheral blood of healthy subjects. The data was obtained from 8 healthy subjects, with a total of 400 samples from 100 microscopic slides with a size of 720×576 pixels. The images were classified by a hematologist into normal leukocytes, including 50 basophils, 19 eosinophils, 29 lymphocytes, 24 monocytes, and 28 neutrophils.

4.2   Settings

The experiment was experimented under four frameworks in semi-supervised learning: SupOnly (trained on the labeled dataset only), ST (trained in a standard self-training manner), ST++ (trained in a self-training++ framework), and SISU (our proposed method). The labeled set accounts for 1/4 of the size of the training set. This proportion is following the setting used in the semi-supervised learning setting, such as ST++ [22].

Without consistency, the batch size for datasets Zheng 1, Zheng 2, and LISC is 16. With the consistency incorporated, the batch size for the labeled and the unlabeled data of Zheng 1 is 16. For Zheng 2 and LISC, the labeled data's batch size is 16 and the unlabeled data's batch size is 8. The network architectures are DeepLabV3+, DeepLabv2, and PSPNet. The choice of the DeepLab networks was because of their atrous convolution which enables wider context absorption. PSPNet also aggregates context at different scales. All of these attributes enable the networks to segment the nucleus and cytoplasm of the cells effectively. Each network architecture consists of the ResNet-50 and ResNet-101 backbone, but DeepLabv2 only has the ResNet-101 backbone. The learning rate for three datasets is 0.009 and follows a poly decay schedule $lr = baselr \times \left(1 - \frac{iter}{total\_iter}\right)^{0.9}$.

Our experiments were recorded using the Mean Intersection over Union (IoU), where mIOU is the area of overlap between the predicted segmentation and the ground truth divided by the area of union between the predicted segmentation and the ground truth.

5   Results

5.1   Performance Comparison Between SISU and Other Approaches

Table 1. Comparison of Performance between SupOnly, ST, ST++, and SISU on three cell segmentation datasets.

Model	Backbone	Datasets											
		Zheng 1				Zheng 2				LISC			
		SupOnly	ST	ST++	SISU	SupOnly	ST	ST++	SISU	SupOnly	ST	ST++	SISU
U-net	R-50	87.39	91.15	92.39	**93.54**	85.74	86.32	87.87	**88.42**	72.28	75.42	75.51	**76.21**
	R-101	86.6	90.27	90.92	**91.37**	86.51	87.2	88.0	**88.92**	74.37	75.68	76.25	**77.02**
DeepLabv3	R-50	87.81	88.04	89.49	**90.28**	84.19	86.02	86.19	**87.81**	73.96	75.33	75.2	**75.62**
	R-101	86.61	88.81	**88.86**	88.78	85.15	87.28	87.03	**88.23**	75.09	75.82	75.95	**76.03**
PSPNet	R-50	79.02	80.44	80.27	**82.18**	74.95	75.09	75.76	**75.92**	69.53	**72.54**	71.17	71.09
	R-101	80.69	81.68	81.56	**82.39**	77.33	76.8	77.76	**78.52**	71.35	72.68	**72.91**	72.75

The results showed in Table 1 indicate a notable enhancement in performance when employing various self-training (ST) schemes compared to the supervised-only (SupOnly) approach. Noticeably, our proposed SISU framework emerges as the most effective method among these alternatives. With the inherent advantage of utilizing a larger dataset, all three methods ST, ST++, and SISU consistently outperform SupOnly, particularly when faced with a scarcity of labeled data. Upon closer examination of ST and ST++ performances, ST++

demonstrates superiority by employing a more fine-tuned approach to categorize pseudo-labeled data based on their reliability. Under the setting of three architectures with 2 different backbones: Resnet-50 and Resnet-101, SISU shows effectiveness when employing the idea of switching supervised training with existent labeled data and unsupervised training with co-training and consistency training. SISU, itself, extends the previous version of ST++, where the approach's limitations are the lack of auxiliary information under various augmentation. SISU achieved state-of-the-art results on most of the settings, except the LISC dataset due to its noisy and out-of-distribution issues.

5.2 Performance on the Supervised Training Phase with the Incorporation of FixMatch

Table 2. Ablation studies of our proposed modules using ResNet-50 as the backbone on Zheng 1, Zheng 2, and LISC datasets.

		Modules			Result (mIOU)		
		Self-Training	FP CVCT	WS Consistency	Zheng 1	Zheng 2	LISC
UNet + R-50	**SupOnly**				87.39	85.74	72.28
	ST++	✓			91.15	86.32	75.42
	Our Proposed Method	✓	✓		91.83	86.88	75.18
		✓		✓	93.21	88.05	**76.43**
		✓	✓	✓	**93.54**	**88.42**	76.21

From ablation studies experiments, Table 2 shows that our strategy is promising, as it achieves state-of-the-art in most experiments on three datasets. Based on the same UNet architecture with backbone ResNet-50, the self-training strategy (ST++) appears to be robust when exploiting information from an unlabeled dataset, improving up to approximately 4% compared to SupOnly. Our design includes two novel modules: Feature Perturbed Cross-view Co-Training (FP CVCT) and Weak-to-Strong Consistency. We conducted thorough ablation investigations on each new approach, in addition to a basic self-training strategy. While FP CVCT appears to allow multiple perspectives on information through feature augmentation, performance across three findings has slightly improved. On the other hand, the robust, yet simple, weak-to-strong consistency outperforms the cell segmentation task. The model improved the most while using the WS Consistency module. The best results for Zheng 1 and Zheng 2 are 93.54% and 88.42%, respectively, under our proposed framework SISU. However, the dataset LISC, which contains diverse information with noisy labeling, attained the best SISU design outcome without the use of FP CVCT. As far as we know, the FP CVCT is sensitive and easily corrupted under various noisy labeling conditions and large out-of-distribution rates.

5.3   Qualitative Results of Prediction Masks from Proposed Method

As depicted in Fig. 2, the proposed SISU approach demonstrates a significant improvement over the previous ST++ method. Using sample image data from the first dataset, Zheng 1, which consists of high-quality, high-contrast images of white blood cells, the SISU approach already generated noticeably better pseudo-masks compared to ST++. In previous research using ST++ for segmenting white blood cell images, even with the application of weak-to-strong consistency, the method struggled to provide accurate pseudo-masks for challenging images, such as those found in the LISC dataset. However, our proposed SISU approach effectively addresses the poor quality of these images, producing pseudo-masks that closely match the ground truth masks. This advancement underscores the robustness and accuracy of the SISU approach in handling diverse image qualities (Fig. 3).

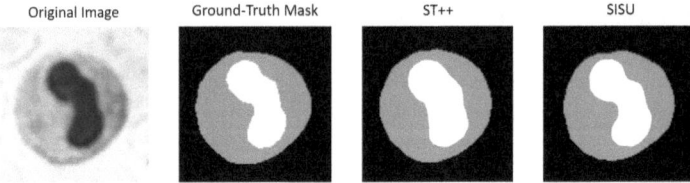

Fig. 2. Easy sample of Ground truth mask, alongside ST++ and SISU masks. The images are produced from the Zheng 1 dataset.

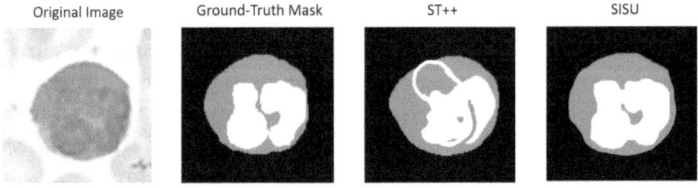

Fig. 3. Hard Example of Ground truth mask, alongside ST++ and SISU masks. The images are produced from the LISC dataset.

6   Conclusion

In our study, we introduce an innovative approach that integrates co-training, and consistency regularization with both traditional and advanced self-training methods. Our findings generally indicate a positive impact from this integration. We observed a compromise between achieving high performance with labeled images and enhancing performance on labeled images through the application of consistency regularization to unlabeled images. Moreover, despite incorporating weak-to-strong consistency and the feature perturbed cross-view co-training, our proposed SISU framework still struggles with noisy pseudo-labeling and confirmation biases. In addition, it is notable that the current results are obtained from only 1/4 of the labeled data, which can limit the generalizability of SISU. To overcome these challenges, our future work will focus on crafting domain-specific strategies tailored to the segmentation of white blood cells and noisy labeling handling techniques for further exploration. One of the concerns is the computational cost and time inference, which will also be carefully considered in our future works. Besides, we plan to conduct comparative experiments using different proportions of labeled data and compare our approach with other existing methods. This will allow us to evaluate the effectiveness and efficiency of our proposed techniques in comparison to alternative approaches.

References

1. Sung, H., et al.: Global cancer statistics 2020: GLOBOCAN estimates of incidence and mortality worldwide for 36 cancers in 185 countries. CA Cancer J. Clin. **71**(3), 209–249 (2021)
2. Roy, R.M., Ameer, P.: Segmentation of leukocyte by semantic segmentation model: a deep learning approach. Biomed. Signal Process. Control **65**, 102385 (2021)
3. Khamael, A.D., Banks, J., Nugyen, K., Al-Sabaawi, A., Tomeo-Reyes, I., Chandran, V.: Segmentation of white blood cell, nucleus and cytoplasm in digital haematology microscope images: a review-challenges, current and future potential techniques. IEEE Rev. Biomed. Eng. **14**, 290–306 (2020)
4. Bispo, J.A.B., Pinheiro, P.S., Kobetz, E.K.: Epidemiology and etiology of leukemia and lymphoma. Cold Spring Harbor Perspect. Med. **10**(6), a034819 (2020)
5. Wilkins, B.S.: Pitfalls in lymphoma pathology: avoiding errors in diagnosis of lymphoid tissues. J. Clin. Pathol. **64**(6), 466–476 (2011)
6. Labati, R. D., Piuri, V., Scotti, F.: All-IDB: the acute lymphoblastic leukemia image database for image processing. In: 2011 18th IEEE International Conference on Image Processing, pp. 2045–2048. IEEE (2011)
7. Mourya, S., Kant, S., Kumar, P., Gupta, A., Gupta, R.: LL Challenge dataset of ISBI 2019 (C-NMC 2019) (Version 1) [dataset]. The Cancer Imaging Archive (2019). https://doi.org/10.7937/tcia.2019.dc64i46r
8. Saleem, S., Amin, J., Sharif, M., Mallah, G.A., Kadry, S., Gandomi, A.H.: Leukemia segmentation and classification: a comprehensive survey. Comput. Biol. Med. **150**, 106028 (2022). https://doi.org/10.1016/j.2022.106028
9. Truong, T.T., Nguyen, H.T., Lam, T.B., Nguyen, D.V., Nguyen, P.H.: Delving into ipsilateral mammogram assessment under multi-view network. In: Cao, X., Xu, X., Rekik, I., Cui, Z., Xi, O. (eds.) MLMI 2023, pp. 367–376. Springer, Cham (2023). https://doi.org/10.1007/978-3-031-45676-3
10. Nguyen, T.H., et al.: Towards robust natural-looking mammography lesion synthesis on ipsilateral dual-views breast cancer analysis. In: Proceedings of the IEEE/CVF International Conference on Computer Vision, pp. 2564–2573 (2023)

11. Ngo, B.H., Lam, B.T., Nguyen, T.H., Dinh, Q.V., Choi, T.J.: dual dynamic consistency regularization for semi-supervised domain adaptation. IEEE Access **12**, 36267–36279 (2024)
12. Le, N.Q.K., Kha, Q.H., Nguyen, V.H., Chen, Y.C., Cheng, S.J., Chen, C.Y.: Machine learning-based radiomics signatures for EGFR and KRAS mutations prediction in non-small-cell lung cancer. Int. J. Mol. Sci. **22**(17), 9254 (2021)
13. Kha, Q.H., Le, V.H., Hung, T.N.K., Le, N.Q.K.: Development and validation of an efficient MRI radiomics signature for improving the predictive performance of 1p/19q co-deletion in lower-grade gliomas. Cancers **13**(21), 5398 (2021)
14. Luu, V.Q., Le, D.K., Nguyen, H.T., Nguyen, M.T., Nguyen, T.T., Dinh, V.Q.: Semi-supervised semantic segmentation using redesigned self-training for white blood cel. arXiv preprint arXiv:2401.07278 (2024)
15. Ngo, T.K.N., et al.: A deep learning-based pipeline for analyzing the influences of interfacial mechanochemical microenvironments on spheroid invasion using differential interference contrast microscopic images. Mater. Today Bio **23**, 100820 (2023)
16. Ronneberger, O., Fischer, P., Brox, T.: U-Net: convolutional networks for biomedical image segmentation. In: Navab, N., Hornegger, J., Wells, W.M., Frangi, A.F. (eds.) MICCAI 2015, Part III. LNCS, vol. 9351, pp. 234–241. Springer, Cham (2015). https://doi.org/10.1007/978-3-319-24574-4_28
17. He, K., Zhang, X., Ren, S., Sun, J.: Deep residual learning for image recognition. In: Proceedings of the IEEE Conference on Computer Vision and Pattern Recognition, pp. 770–778 (2016)
18. Sohn, K., e al.: FixMatch: simplifying semi-supervised learning with consistency and confidence. Adv. Neural. Inf. Process. Syst. **33**, 596–608 (2020)
19. Cascante-Bonilla, P., Tan, F., Qi, Y., Ordonez, V.: Curriculum labeling: revisiting pseudo-labeling for semi-supervised learning. In: Proceedings of the AAAI Conference on Artificial Intelligence, vol. 35, no. 8, pp. 6912–6920 (2021)
20. Zheng, X., Wang, Y., Wang, G., Liu, J.: Fast and robust segmentation of white blood cell images by self-supervised learning. Micron **107**, 55–71 (2018)
21. Rezatofighi, S.H., Soltanian-Zadeh, H.: Automatic recognition of five types of white blood cells in peripheral blood. Comput. Med. Imaging Graph. **35**(4), 333–343 (2011)
22. Yang, L., Zhuo, W., Qi, L., Shi, Y., Gao, Y.: St++: make self-training work better for semi-supervised semantic segmentation. In: Proceedings of the IEEE/CVF Conference on Computer Vision and Pattern Recognition, pp. 4268–4277 (2022)

Optimizing Foundation Models for Histopathology: A Continual Learning Approach to Cancer Detection

Ankur Yadav[⊠][iD] and Ovidiu Daescu[iD]

The University of Texas at Dallas, Richardson, TX 75080, USA
ankur.yadav@utdallas.edu

Abstract. Accurate analysis of histopathology images is a key step in cancer diagnosis and treatment planning. Specialized computational models for histopathology image analysis may be needed to better capture the unique characteristics of histopathological images that could be missed by models pretrained on generic data, such as ImageNet. We employed an incremental learning approach to enhance the performance of foundation models using the EfficientNet B0 architecture. Our study comprised three key experiments. First, we established a baseline accuracy of 84.4% by training the model on 50% of our dataset. Second, we investigated the impact of retraining different numbers of top blocks during incremental learning phases, finding that retraining three to six blocks provided the most significant accuracy improvements. Third, we analyzed the trade-off between accuracy improvement and training time, determining that retraining three to six blocks offered the best balance between performance gains and computational efficiency. Our results demonstrate the effectiveness of specialized models in capturing the details specific of histopathological images. Despite computational limitations, our findings underscore the importance of tailored histopathological models.

Keywords: Histopathology · Machine Learning · Transfer Learning · Continual Learning

1 Introduction

The microscopic examination of tissue samples in histopathology is essential for diagnosing and treatment planning for most cancer types. Traditional histopathological analysis, relying on pathologists to manually examine tissue or digital slides, is time-consuming, prone to human error, and limited by the availability of skilled professionals. With the increasing incidence of cancer, there is a critical need for automated, accurate diagnostic tools to aid pathologists. Automated diagnosis using advanced computational techniques can enhance efficiency, accuracy, and consistency, possibly leading to better patient outcomes and optimizing medical resources. This study explores the potential of digital histopathology image based deep learning models optimized through incremental learning to meet the growing demands for automated diagnostic solutions in histopathology.

H. Chen et al. (Eds.): TAI4H 2024, LNCS 14812, pp. 145–156, 2024.
https://doi.org/10.1007/978-3-031-67751-9_12

Machine learning and artificial intelligence have significantly impacted histopathology, providing new methods for the automated analysis and classification of histopathological images. Wu et al. (2022) highlighted the transformative potential of deep learning for computational histopathology, reviewing techniques such as color normalization, nuclei segmentation, and cancer diagnosis and prognosis [14]. Yadav et al. (2023) demonstrated the efficacy of topological signatures for cancer detection, using advanced computational methods to enhance diagnostic accuracy [15]. Kumar et al. (2024) introduced ML3CNet, an automatic framework for lung cancer subtype classification using non-local means-assisted methods, showcasing sophisticated machine learning techniques for accurately classifying lung cancer subtypes [5]. These studies underscore the significant advancements and potential of computational models in enhancing histopathology based diagnostics.

Recent advancements in transfer learning have led to the development of several effective models for image recognition tasks, such as VGG, InceptionV3, and EfficientNet, which have shown remarkable performance when applied to histopathology. The VGG model, introduced by Simonyan and Zisserman (2014), improves image classification accuracy using deep convolutional networks with small receptive fields [11]. InceptionV3, proposed by Szegedy et al. (2016), balances depth and width, incorporating factorized convolutions and aggressive regularization techniques to enhance pattern recognition in images [12]. EfficientNet, developed by Tan and Le (2019), uses a compound scaling method to uniformly scale network dimensions, achieving state-of-the-art accuracy with better efficiency [13]. When employed in transfer learning, these models leverage their pretrained weights on large scale image datasets to adapt effectively to specific tasks in histopathology, demonstrating their robustness and versatility.

Research has shown the effectiveness of transfer learning models in histopathology. Yadav et al. (2023) used transfer learning to classify subtypes of rhabdomyosarcoma in whole slide images, improving performance on limited data samples [16]. Arooj et al. (2022) applied transfer learning for breast cancer detection, significantly improving diagnostic accuracy by fine-tuning pretrained models with histopathological images [2]. Mahmud et al. (2023) conducted a deep analysis of transfer learning-based breast cancer detection, experimenting with various pretrained models to determine the most effective approach for classifying breast cancer sub-types [6]. Nasir et al. (2022) integrated transfer learning with blockchain, fog computing, and edge computing technologies for osteosarcoma detection, achieving high accuracy while ensuring data security and processing efficiency [7]. These studies collectively highlight the advancements and potential of transfer learning in enhancing histopathological diagnostics.

Despite these advancements, the need for new transfer learning models tailored explicitly for histopathology images remains critical. Current models pretrained on datasets like ImageNet may not fully capture the subtle characteristics of histopathological images, such as intricate cellular patterns and staining variations. Developing specialized models that can leverage the nuanced features of histopathology images is essential for improving diagnostic reliability (Fig. 1).

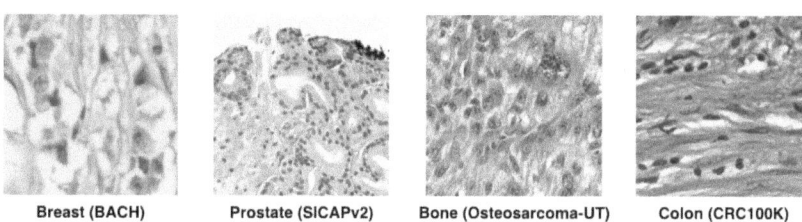

Breast (BACH)   **Prostate (SICAPv2)**   **Bone (Osteosarcoma-UT)**   **Colon (CRC100K)**

Fig. 1. Representative histopathological image samples from different cancer datasets used in this study. From left to right: Breast tissue from the BACH dataset, prostate tissue from the SICAPv2 dataset, bone tissue from the UT-Osteosarcoma dataset, and colon tissue from the CRC100K dataset. These images illustrate the diverse morphological characteristics and staining patterns analyzed by our model.

Incremental learning allows models to learn from new data without forgetting previously acquired knowledge, which is of great importance, given the dynamic and ever growing nature of medical datasets. Integrating new information while retaining past knowledge ensures that models remain up to date and relevant, providing consistent performance.

In conclusion, while significant progress has been made with transfer learning in histopathology, there is a need for models specifically designed and trained for this domain. Coupled with incremental learning strategies, these models have the potential to offer substantial advancements in the automated analysis of histopathological images, leading to more accurate and efficient diagnostic tools in clinical practice. Our work contributes to addressing this problem by leveraging incremental learning to enhance model performance and adaptability, providing a foundation for developing specialized histopathology models.

2   Materials and Methods

2.1   Dataset Description

In this study, we used publicly available benchmark datasets to train and validate our models for histopathology image analysis. These datasets encompass various types of cancer and healthy tissues, providing a diverse set of samples for classification tasks. They are collectively used to classify images into tumor and healthy tiles. The datasets included in this study are:

- **BACH (Breast Cancer Histology Images)** [1]:
    - *Description*: The BACH dataset, presented at the ICIAR 2018 Grand Challenge, includes a diverse range of histopathological images indicative of various stages of breast cancer. The images are categorized into four types: normal, benign, in situ carcinoma, and invasive carcinoma.
    - *Dataset Size*: The dataset contains 46,535 images, with 32,347 healthy tissue and 14,188 images of tumor tissue.

Table 1. Distribution of image samples across different cancer datasets for training and testing.

Dataset	Class	Number of Images
BACH	Healthy	32,347
	Tumor	14,188
SICAPv2	Healthy	9,523
	Tumor	25,408
UT-Osteosarcoma	Healthy	5,982
	Tumor	16,324
CRC100K	Healthy	6,134
	Tumor	27,390

- **SICAPv2 (Prostate Cancer Dataset)** [10]:
  - *Description*: SICAPv2 contains annotated prostate histology Whole Slide Images (WSIs) with detailed global Gleason scores and patch-level Gleason grades. The dataset includes images of normal prostate tissue as well as various grades of cancerous tissue.
  - *Dataset Size*: The dataset comprises 34,931 images, with 9,523 images of healthy tissue and 25,408 images of tumor tissue.
- **UT-Osteosarcoma (Bone Cancer Dataset)** [3]:
  - *Description*: This dataset offers a collection of Hematoxylin and Eosin (H&E) stained osteosarcoma histology images from the UTSW Medical Center. It includes images of healthy bone tissue and tissue affected by osteosarcoma.
  - *Dataset Size*: The UT-Osteosarcoma dataset includes 22,306 images, with 5,982 healthy tissue and 16,324 images of tumor tissue (Table 1).
- **CRC100K (Colorectal Cancer Dataset)** [4]:
  - *Description*: CRC100K features 100,000 histological images of colorectal cancer and healthy tissue. The images are categorized into normal, hyperplastic, adenoma, in situ carcinoma, and invasive carcinoma.
  - *Dataset Size*: The dataset consists of 33,524 images, with 6,134 healthy tissue and 27,390 images of tumor tissue.

By integrating these datasets, we aim to develop a robust model capable of accurately distinguishing between tumor and healthy tissue across different types of cancer. The combined dataset provides a diverse representation of histopathological images, facilitating the creation of a versatile diagnostic tool. This approach ensures the model is exposed to various tissue types and staining patterns, enhancing its ability to generalize across different medical datasets.

2.2   Data Preprocessing

Effective machine learning for image-based analysis depends on the quality of the training data. We employed a detailed preprocessing protocol to prepare histopathological images for deep learning:

1. **Tile Extraction**: Large images were divided into smaller, non-overlapping tiles labeled as healthy or tumor, increasing the number of training samples and ensuring the model learns localized features of cancerous tissue.
2. **Resizing**: Images were resized to 224×224 pixels to balance detection of histological patterns with computational efficiency, and ensuring uniformity across datasets.
3. **Normalization**: To adjust for variations in staining techniques, each image underwent a rigorous normalization process, scaling pixel values to the range $[0, 1]$ to focus on morphological features rather than stain discrepancies.
4. **Augmentation**: Techniques such as random rotations, flips, and scalings were used to expand the dataset, allowing to better capture features that are invariant to such transformation and enhancing robustness and generalization.

2.3   Model Training and Evaluation

We initially used 50% of the dataset for training, divided into 80% for training and 20% for evaluation. With each iteration, we added 10% of the dataset incrementally, maintaining an 80–20 split between training and evaluation subsets. This method ensured the model progressively learned from a growing dataset while validating performance on a representative evaluation subset, facilitating both learning and generalization.

3   Experimental Setup

This section describes the hardware and software configurations, the foundation model architecture, the rationale behind selecting EfficientNet B0, and the training protocols used in our study.

3.1   Hardware Configuration

The experiments were conducted using a computing setup that included the following hardware:

- **GPUs**: NVIDIA RTX 2060 Super, with 8 GB of VRAM. These GPUs provide the necessary computational power for training deep learning models on sizeable histopathological image datasets.
- **RAM**: 16 GB, to ensure efficient data processing and multitasking during model training and evaluation.

3.2   Software Configuration

The software environment was configured with the following tools and libraries:

- **Framework**: PyTorch [8], selected for its flexibility and extensive support for advanced modeling techniques (Fig. 2).

- **Libraries**: NumPy, SciPy, and scikit-learn [9], used for effective data manipulation, statistical analysis, and classical machine learning operations. This suite of tools provides a robust environment for preprocessing data and executing complex model training algorithms.

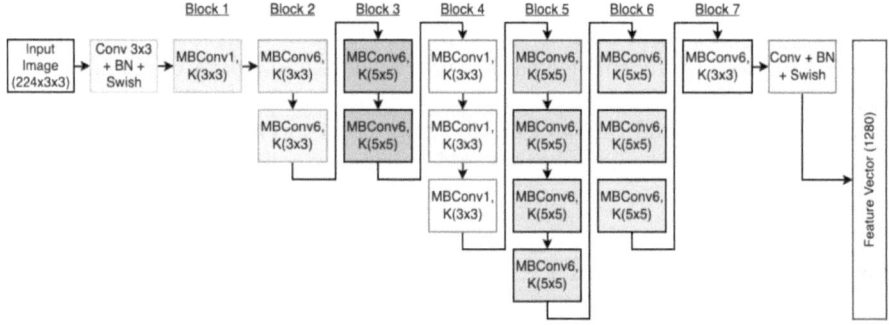

Fig. 2. Architecture of EfficientNet B0, illustrating the convolutional layers and mobile inverted bottleneck convolution (MBConv) blocks used for feature extraction. The network begins with a standard convolutional layer and seven blocks of varying configurations. Each block consists of MBConv layers with kernel sizes denoted as $K(3 \times 3)$ or $K(5 \times 5)$. The final feature vector has a dimensionality of 1280, extracted after a concluding convolutional layer with batch normalization and Swish activation.

3.3   Foundation Model Architecture

The foundation model architecture utilized in this study is EfficientNet B0. EfficientNet B0 is part of the EfficientNet family, which employs a compound scaling method that uniformly scales all dimensions of depth, width, and resolution using a set of fixed scaling coefficients. This architecture is designed to achieve high performance while maintaining computational efficiency.

- **Initial Layers**: The model begins with a standard convolutional layer.
- **Intermediate Layers**: A series of mobile inverted bottleneck convolutions (MBConv) with kernel sizes $K(3 \times 3)$ or $K(5 \times 5)$.
- **Final Layers**: The final feature vector has a dimensionality of 1280, extracted post a concluding convolutional layer with batch normalization and Swish activation.

3.4   Rationale for Selecting EfficientNet B0

EfficientNet B0 was selected for this study due to its balance between high performance and computational efficiency. The specific reasons for choosing this architecture include:

- **Performance and Complexity**: EfficientNet B0 provides state-of-the-art accuracy on the ImageNet dataset with comparatively lower complexity (only 5.3M parameters) and computational demand.
- **Dataset Size Consideration**: The architecture's lower complexity is advantageous for our dataset, which is much smaller than ImageNet, mitigating the risk of overfitting while still capitalizing on the model's robust feature extraction abilities.
- **Computational Feasibility**: Given the hardware limitations and the need for iterative testing, EfficientNet B0's design, requiring fewer computational resources, facilitates broader experimentation without sacrificing training and evaluation efficiency.
- **Rationale for Retraining**: Retraining the model on histopathological images, rather than using it as pre-trained on ImageNet, customizes its feature extraction to focus on cancer-specific characteristics.

3.5   Training Protocols

The methodology behind training our models is pivotal to our study's aim of enhancing histopathological image classification. The following protocols were employed:

1. **Initial Training**: The entire EfficientNet B0 model was trained using 50% of the dataset to establish a robust baseline for feature extraction capabilities.
2. **Incremental Training**: Subsequently, the model was incrementally trained in phases, each using an additional 10% of the dataset. Different combinations of blocks of the EfficientNet B0 architecture during these phases were retrained to refine the model's higher-level feature representations.
3. **Optimization**: Model optimization was conducted using the Adam optimizer. Early stopping based on validation loss was implemented to avoid overfitting, ensuring the model maintains high generalization performance.
4. **Evaluation Metric**: The effectiveness of the model was assessed using its accuracy as the main measure.

These protocols underscore our commitment to developing and validating models that significantly improve the accuracy and efficiency of detecting and classifying cancer types in histopathological images.

4   Experiments and Results

In this section we detail our experimental assessment of the proposed model.

4.1   Experiment 1: Baseline Training

The first experiment aimed to establish a baseline performance for the EfficientNet B0 model by training it with 50% of the available dataset. This initial training phase was essential for understanding the model's capabilities and setting a benchmark for subsequent incremental learning phases.

The dataset was randomly divided, with 50% allocated for initial training and the remaining 50% reserved for incremental training phases and validation. The model was trained using standard training protocols, with the Adam optimizer and a learning rate of 1×10^{-3}. Early stopping based on validation loss was implemented to prevent overfitting.

The baseline training of the EfficientNet B0 model on 50% of the dataset yielded an accuracy of **84.4%**. This performance provided a baseline for evaluating the effectiveness of the incremental learning strategy implemented in subsequent experiments.

The results from this baseline training phase indicated that the EfficientNet B0 model could achieve a reasonable level of accuracy when trained on a significant subset of the data. This experiment set the foundation for the subsequent incremental training experiments, which aimed to improve the model's performance by gradually introducing additional data.

4.2 Experiment 2: Incremental Training with Varying Number of Layers

In the second experiment, we aimed to determine the optimal number of layers to be retrained during incremental learning phases of the EfficientNet B0 model. We incrementally trained the model with an additional 10% of data in each phase, conducting seven experiments, varying the number of top blocks retrained while keeping the rest frozen.

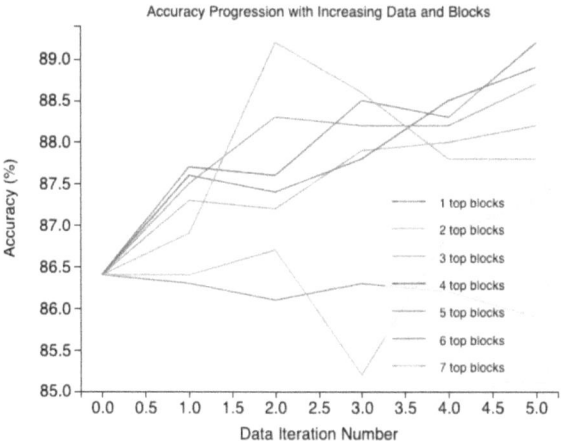

Fig. 3. Accuracy progression with increasing data and blocks. The plot shows the accuracy achieved by retraining different numbers of top blocks in the EfficientNet B0 model with each 10% data increment.

To balance new data integration and avoid catastrophic forgetting, we implemented a two-phase training strategy:

1. 5 epochs on the new 10% of data.
2. 5 epochs on the combined dataset (initial 50% + newly added data).

The results, illustrated in Fig. 3, show varying improvements in accuracy depending on the number of top blocks retrained. Retraining three to six top blocks consistently demonstrated stable and positive accuracy trends, indicating an effective balance between learning new features and retaining previously learned information. Configurations with fewer blocks exhibited more volatility, while retraining all seven blocks initially showed performance dips, suggesting overfitting that stabilized with further data integration.

This experiment highlights the importance of selecting an appropriate number of layers for retraining to maximize performance while minimizing the computational expense and overfitting risks.

4.3   Experiment 3: Trade-Off Between Accuracy Improvement and Training Time

This experiment aimed to understand the trade-off between accuracy improvement and training time required for retraining different numbers of blocks.

Methodology

- **Baseline Training**: The model was initially trained on 50% of the dataset.
- **Incremental Training Phases**: Incrementally trained with an additional 10% of data at each iteration, similar to the previous experiment, with each phase consisting of 10 epochs.
- **Block Configuration**: Retraining from 1 to 7 top blocks; for each configuration, average accuracy improvement was calculated from five incremental iterations, and training time was logged in hours.

Results: Figure 4 illustrates the trade-off between average accuracy improvement and training time for different numbers of retrained blocks.

- **Accuracy Improvement**: Retraining three to six top blocks provided the highest accuracy improvements, balancing new feature learning and retention of previous knowledge.
- **Training Time**: Training time increased with the number of retrained blocks. Fewer blocks required less time but showed smaller accuracy improvements.
- **Optimal Configuration**: Retraining three to six top blocks offered significant accuracy gains with reasonable training time. Retraining all seven blocks provided higher accuracy but at a substantially increased training time, indicating diminishing returns.

This experiment highlighted the importance of selecting an appropriate number of layers for retraining to maximize performance gains while effectively managing computational resources.

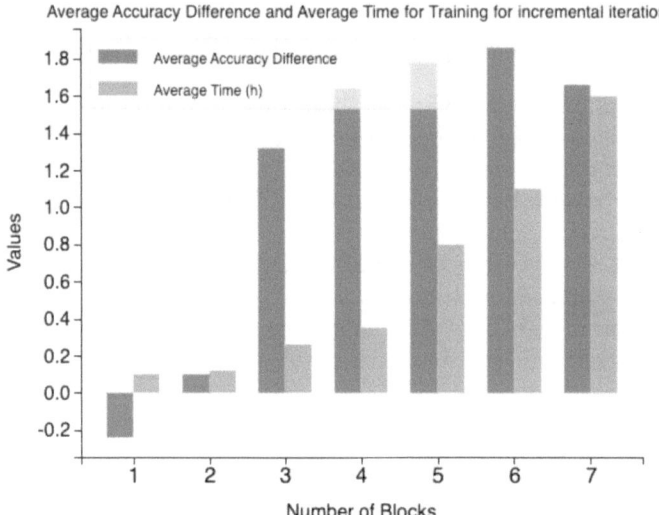

Fig. 4. Trade-off between average accuracy improvement and training time for different numbers of retrained blocks.

5 Discussion

In this study, we assessed a specialized model for histopathology image analysis, pre-trained only on histopathology images. Histopathological images exhibit unique characteristics and patterns that require tailored feature extraction methods. Our experiments indicated that a model explicitly trained on histopathology data could potentially outperform models that rely on general-purpose pretrained models. The study provides a foundation for further research into developing and optimizing models tailored for histopathology.

We established a baseline in the first experiment by training the EfficientNet B0 model on 50% of our dataset. The initial accuracy of **84.4%** highlighted the potential of this architecture but also underscored the need for further optimization through incremental learning. This initial phase established refined training strategies to enhance the performance of the model.

The second experiment explored the impact of retraining different numbers of top blocks during the incremental learning phases. We found configurations involving three to six top blocks retraining provided the most significant accuracy improvements. Notably, the best accuracy of **89.2%** was achieved when retraining six blocks with 100% of the data. This experiment demonstrated the importance of balancing new feature learning with model retaining. The results indicated that selectively retraining particular layers could enhance the adaptability and accuracy of the model.

The third experiment focused on the trade-off between accuracy improvement and training time. By averaging the accuracy gains and training times across multiple incremental iterations, we identified that retraining three to six top blocks offered the best compromise between performance and computational efficiency. Notably, retraining six blocks yielded the highest accuracy improvement of **1.86%**, taking **1 h and 6 min** for retraining. However, considering the trade-off between accuracy improvement and training time, retraining four blocks provided the most efficient balance, as it achieved significant accuracy gains with the least time expenditure relative to other configurations. This finding emphasizes the importance of balancing accuracy improvements with computational efficiency for practical applications.

To validate our trained feature extractor, we tested it on external datasets and observed notable performance improvements. These results underscore the effectiveness of our specialized model in capturing the intricate details present in histopathological images, which could be missed by general-purpose models. Although a detailed comparative analysis with other foundation models like EfficientNet, InceptionV3, VGG16, and VGG19 is beyond the scope of this paper, our findings suggest that specialized models tailored for histopathology have a significant potential to outperform general-purpose models. Future work should include comprehensive comparative studies to substantiate these findings further.

Despite the promising results, our study faced several challenges and limitations. The primary constraint was the limited computational power available, necessitating using EfficientNet B0. While this model provided a good starting point, larger datasets and more complex models could yield even better results. Future work should explore the scalability of our approach with more robust architectures and extensive datasets to enhance diagnostic accuracy and reliability further.

In conclusion, our study highlights the effectiveness of developing specialized models for histopathology image analysis. The incremental learning approach improved model performance while efficiently balancing computational resources. Future research should focus on leveraging more advanced architectures and larger datasets to push the boundaries of what is achievable in this critical domain.

References

1. Aresta, G., et al.: BACH: grand challenge on breast cancer histology images. Med. Image Anal. **56**, 122–139 (2019)
2. Arooj, S., Zubair, M., Khan, M.F., Alissa, K., Khan, M.A., Mosavi, A.: Breast cancer detection and classification empowered with transfer learning. Front. Public Health **10**, 924432 (2022)
3. Arunachalam, H.B., et al.: Viable and necrotic tumor assessment from WSI of osteosarcoma using ML and DL models. PLoS ONE **14**(4), e0210706 (2019)
4. Kather, J.N., Halama, N., Marx, A.: 100,000 histological images of human colorectal cancer and healthy tissue. Zenodo10 **5281** (2018)
5. Kumar, A., Vishwakarma, A., Bajaj, V.: ML3CNet: non-local means-assisted automatic framework for lung cancer subtypes classification using histopathological images. Comput. Methods Programs Biomed. **251**, 108207 (2024)

6. Mahmud, M.I., Mamun, M., Abdelgawad, A.: A deep analysis of transfer learning based breast cancer detection using histopathology images. In: 2023 10th International Conference on Signal Processing and Integrated Networks (SPIN), pp. 198–204. IEEE (2023)

7. Nasir, M.U., Khan, S., Mehmood, S., Khan, M.A., Rahman, A.U., Hwang, S.O.: IoMT-based osteosarcoma cancer detection in histopathology images using transfer learning empowered with blockchain, fog computing, and edge computing. Sensors **22**(14), 5444 (2022)

8. Paszke, A., et al.: PyTorch: an imperative style, high-performance deep learning library. In: Proceedings of the 33rd International Conference on Neural Information Processing Systems (NeurIPS) (2019)

9. Pedregosa, F., et al.: Scikit-learn: ML in Python. JMLR **12**, 2825–2830 (2011)

10. Silva-Rodríguez, J., Colomer, A., Sales, M.A., Molina, R., Naranjo, V.: Going deeper through the Gleason scoring scale: an automatic end-to-end system for histology prostate grading and cribriform pattern detection. Comput. Methods Programs Biomed. **195**, 105637 (2020)

11. Simonyan, K., Zisserman, A.: Very deep convolutional networks for large-scale image recognition. arXiv preprint arXiv:1409.1556 (2014)

12. Szegedy, C., Vanhoucke, V., Ioffe, S., Shlens, J., Wojna, Z.: Rethinking the inception architecture for computer vision. In: Proceedings of the IEEE Conference on Computer Vision and Pattern Recognition (2016)

13. Tan, M., Le, Q.V.: EfficientNet: rethinking model scaling for convolutional neural networks. In: Proceedings of the 36th International Conference on Machine Learning (ICML), pp. 6105–6114 (2019)

14. Wu, Y., et al.: Recent advances of deep learning for computational histopathology: principles and applications. Cancers **14**(5), 1199 (2022)

15. Yadav, A., Ahmed, F., Daescu, O., Gedik, R., Coskunuzer, B.: Histopathological cancer detection with topological signatures. In: 2023 IEEE International Conference on Bioinformatics and Biomedicine (BIBM), pp. 1610–1619. IEEE (2023)

16. Yadav, A., Daescu, O., Leavey, P., Rudzinski, E.: Machine learning for rhabdomyosarcoma whole slide images sub-type classification. In: Proceedings of the 16th International Conference on PErvasive Technologies Related to Assistive Environments, pp. 192–196 (2023)

RF-Lung-DR: Integrating Biological and Drug SMILES Features in a Random Forest-Based Drug Response Predictor for Lung Cancer Cell Lines

Thi-Oanh Tran[1,2], Quang-Hien Kha[2,3], and Nguyen Quoc Khanh Le[2,4,5(✉)] ⓘ

[1] International Ph.D. Program in Cell Therapy and Regenerative Medicine, College of Medicine, Taipei Medical University, Taipei 11031, Taiwan
d151110003@tmu.edu.tw

[2] AIBioMed Research Group, Taipei Medical University, Taipei 11031, Taiwan
d142111015@tmu.edu.tw

[3] International Ph.D. Program in Medicine, Taipei Medical University, Taipei 11031, Taiwan

[4] Professional Master Program in Artificial Intelligence in Medicine, College of Medicine, Taipei Medical University, Taipei 11031, Taiwan
khanhlee@tmu.edu.tw

[5] Research Center for Artificial Intelligence in Medicine, Taipei Medical University, Taipei 11031, Taiwan

Abstract. In the era of precision medicine, predicting drug responses accurately is crucial for tailoring patient-specific treatments. Despite advances in machine learning (ML) models for drug response prediction (DRP), challenges remain in predicting effective therapies with high accuracy. This study introduces RF-Lung-DR, a ML model that integrates biological markers and drug SMILES features to predict drug responses in lung cancer cell lines, thus enhancing drug screening processes. Using drug sensitivity data from the Genomics of Drug Sensitivity in Cancer (GDSC), the model was developed across seven ML algorithms, with Random Forest (RF) proving to be the most effective for optimizing DRP accuracy. RF-Lung-DR achieved prediction accuracies of 80% in lung squamous cell carcinoma (LUSC) and 78% in lung adenocarcinoma (LUAD). The investigation also identified key biological biomarkers and drug SMILES features that significantly influence predictive performance. Focusing on lung cancer-a leading cause of cancer-related mortality worldwide-RF-Lung-DR's methodology supports the broader application of personalized medicine and underscores the potential for developing individualized patient care strategies in oncology.

Keywords: Machine Learning · Precision Medicine · Drug Sensitivity · Personalized treatment · Pharmacogenomics

© The Author(s), under exclusive license to Springer Nature Switzerland AG 2024
H. Chen et al. (Eds.): TAI4H 2024, LNCS 14812, pp. 157–167, 2024.
https://doi.org/10.1007/978-3-031-67751-9_13

1   Introduction

In the realm of contemporary precision medicine, drug response prediction (DRP) has emerged as a crucial endeavor. Moving away from traditional disease-centric approaches that once dictated treatment strategies, the current paradigm advocates for personalized, data-driven methodologies. This transformative shift is propelled by the convergence of omics and big data [7]. In cancer treatment, the utilization of multi-omics data has enriched the development of advanced machine learning (ML) models tailored for individualized treatment, as discussed in recent studies by He et al. [12] and Tran et al. [26]. Coupled with the growth of preclinical models and enhancements in computational techniques for predicting drug responses, pharmacogenomics has seen remarkable progress over the past two decades [9]. Its success is now increasingly recognized in personalized cancer therapy, where it has become common practice in clinical settings.

Lung cancer remains the second most frequently diagnosed cancer in both males and females and continues to be the leading cause of cancer-related deaths globally in 2024, according to statistics from the American Cancer Society [21, 22]. Despite significant advancements in diagnosis and treatment, the five-year survival rate for lung cancer patients remains modest, hovering around 20% [19]. This is considerably lower compared to the survival rates for other prevalent cancers, such as prostate cancer, the most common type of cancer in males, with over 90% survival rate [1], and breast cancer, the most common type of cancer in women, which has a survival rate of approximately 90% in the US [2]. Therefore, tailoring effective treatment strategies is crucial for improving outcomes in lung cancer treatment.

In recent years, ML-based DRP models have been developed to enhance lung cancer treatment across four main medical applications: predicting patient responses to monotherapies such as EGFR-TKIs [18]; responses to immune checkpoint inhibitors (PD-L1) [29]; repurposing drugs, such as the deep learning model that repurposed pimozide from an anti-dyskinesia agent to an anti-NSCLC drug [16]; and predicting patients' responses to combination therapies using bulk-level drug response RNA-seq information and single-cell sequencing data [5]. However, the application of ML in scanning for lung cancer treatments is limited by challenges such as high-quality data and model interpretability.

Recently, with the expansion of pre-clinical models, pharmacogenomics has achieved considerable success in DRP tasks [9]. Pharmacogenetics, which provides extensive data on drug efficacy, enables the ranking of different drugs' effectiveness on the same cell line [24]. The presence of large-scale pharmacogenomics data sources such as the Genomics of Drug Sensitivity in Cancer (GDSC) [31], PharmacoDB [8] offers a variety of material for therapeutic predictions. In addition, ML algorithms are pivotal in examining large-scale genetic variation datasets, revealing drug response patterns not easily discernible through conventional statistical methods.

In this study, we collected high-quality pharmacogenomics data from the GDSC database and extracted drug SMILES (Simplified Molecular Input Line Entry System) features to develop an effective DRP model for lung cancer cell lines. Drug SMILES, which encode the structural information of molecules in a textual format [3], have shown significant progress and present a novel approach for advancing DRP models [11,32]. We constructed our model guided by five primary objectives: (1) extracting drug SMILES features, (2) identifying the most effective ML classifier, (3) utilizing feature selection techniques to enhance model accuracy, (4) optimizing the DRP model, and (5) investigating the most influential biological and drug features on model performance for clinical applications.

2   Materials and Methods

2.1   Data Collection and Processing

The drug response data for lung cancer cell lines were sourced from the GDSC database (https://www.cancerrxgene.org/), which is the most extensive public repository for information on drug sensitivity in cancer cells and molecular indicators of drug response [31]. We collected two specific drug sensitivity datasets: one for lung adenocarcinoma (LUAD), which included 62 cell lines tested against 288 drugs that generated 15747 drug sensitivities, and another for lung squamous cell carcinoma (LUSC), which comprised 14 cell lines tested against the same number of medications generated 3891 sensitivities. Genomics profiles were provided for all 76 lung cancer cell lines included in the analysis.

We successfully gathered 249 drug SMILES from three sources: DrugBank (https://go.drugbank.com/) [30], PubChem (https://pubchem.ncbi.nlm.nih.gov/) [14], and MedChemExpress (https://www.medchemexpress.com/) [4]. Utilizing RDKit 1.0 [4] and PyBioMed 1.0 [6], employing methods referenced from our laboratory's previous work [13]. In total, eight types of descriptors were used, including: (1) E-State VSA (Valence State Atom) descriptors: those encode electronic and steric properties; (2) PEOE-VSA (Partial Equalization of Orbital Electronegativity - Van der Waals Surface Area), these are based on the partial charges computed using the PEOE method combined with the van der Waals surface area; (3) MTPSA (Molecular Topological Polar Surface Area), this descriptor estimates the ability of a molecule to interact with polar solvents, which is critical for understanding its pharmacokinetics properties such as absorption, distribution, metabolism, and excretion (ADME); (4) VSA/EState (Valence State Atom/Extended Topological State), these descriptors combine electronic and steric information based on the E-State indices and Van der Waals surface areas; (5) LabuteASA (Labute's Approximate Surface Area), this descriptor estimates the approximate surface area of a molecule, which is useful for understanding its interactions with other molecules, solvents, and biological targets; (6) MRVSA (Molecular Refractivity - Van der Waals Surface Area), these descriptors combine molecular refractivity properties with the van der Waals surface area; (7) TPSA (Topological Polar Surface Area), this descriptor represents the surface area of polar atoms (usually oxygen and nitrogen) in a molecule, including their attached hydrogens; (8) slogP-VSA, this descriptor represents the sum of the van der Waals surface areas of the molecule's atoms, weighted by their contributions to the SLogP (octanol-water partition coefficient). We finally extracted 61 useful features for 249 drug SMILES strings which can be used in our ML model.

For labeling drug responses, sensitive cell lines and drugs were annotated as one, while resistant cell lines were annotated as zero. The cutoff for determining sensitivity or resistance was set at the median Area Under the Curve (AUC) values among all cell lines, as referenced from the work of Chen et al. [5].

We finally created two datasets for ML used: (1) the LUAD-DR dataset included interactions of 62 cell lines with 249 drugs resulting in 15,747 drug responses and 149 features (a combination of 88 LUAD biological markers with 61 drug-extracted features); (2) the LUSC-DR dataset encompassed interactions of 14 cell lines with 249 drugs, generating 3,891 drug sensitivities and 134 features (a combination of 73 LUSC biological markers with 61 drug features).

2.2 Model Constructing

Our model development process involved the utilization of seven prevalent ML classifiers: Random Forest (RF), AdaBoost (AB), Support Vector Machine (SVM), Decision Tree (DT), k-Nearest Neighbors (KNN), Naïve Bayes (NB), and Logistic Regression (LR). Firstly, to find the best classifier, two datasets were partitioned into subsets with a 75% to 25% split ratio for training and testing respectively. The classifier demonstrating superior performance on the testing sets was chosen for further refinement.

Enhancements to our model were achieved through a systematic feature selection process. We employed Recursive Feature Elimination with Cross-Validation (RFECV) [17], which iteratively removed less significant features based on the model's performance, thus optimizing the feature set. Additionally, we incorporated cross-validated Ridge regression (RidgeCV) techniques to further refine the feature selection and improve classification accuracy. In this step, we implemented a ten-fold cross-validation (CV) strategy to evaluate model performance. This approach ensured that the evaluation metrics were robust and less prone to overfitting. The classifier demonstrating superior performance on the testing sets was chosen for further refinement.

The final model optimization was carried out by determining the optimal feature group and the most effective classifier. This was achieved by evaluating the performance of each model using a comprehensive ten-fold CV approach. This additional layer of validation provided a more reliable assessment of the model's generalization capability and robustness.

2.3 Evaluation Metrics

To assess the performance of our model, we utilized four key metrics: Accuracy, Precision, Recall, and F1-score [13]. The definitions and calculations are as follows:

$$Accuracy = \frac{TP + TN}{TP + TN + FP + FN} \tag{1}$$

$$Recall = \frac{TP}{TP + FN} \tag{2}$$

$$Precision = \frac{TP}{TP + FP} \tag{3}$$

$$F1 = 2 * \frac{Precision * Recall}{Precision + Recall} \tag{4}$$

In this study, TP (True Positive) refers to correctly identifying a cell line as drug-sensitive, TN (True Negative) to correctly identifying a cell line as drug-resistant, FP (False Positive) to incorrectly identifying a drug-resistant cell line as drug-sensitive, and FN (False Negative) to incorrectly identifying a drug-sensitive cell line as drug-resistant. Accuracy measures how well the model can correctly classify the cell lines as either sensitive or resistant to the drugs. Recall reflects the model's effectiveness in identifying truly sensitive cell lines, highlighting its ability to capture positive cases. Precision indicates the model's reliability in predicting positive (sensitive) cases and its ability to minimize false positives. The F1 score provides a comprehensive measure of the model's accuracy in identifying drug-sensitive and drug-resistant cell lines, taking into account both the model's precision and recall.

3   Results

The framework of our ML-based DRP model for lung cancer cell lines, as illustrated in Fig. 1, consists of four main steps: (A) Data collection and processing: Drug sensitivity data is collected from GDSC2, and drug SMILES are sourced from DrugBank, MedChemExpress, and PubChem. Features are extracted from 249 SMILES using RDKit 1.0 and PyBioMed 1.0. (B) Modeling: The data is divided into training (75%) and testing (25%) sets, using seven common ML algorithms to identify the most effective algorithm. (C) Feature selection: RFECV and RidgeCV techniques are applied to determine the most effective features for the model, and a ten-fold CV was used to fine-tune the model. (D) Model

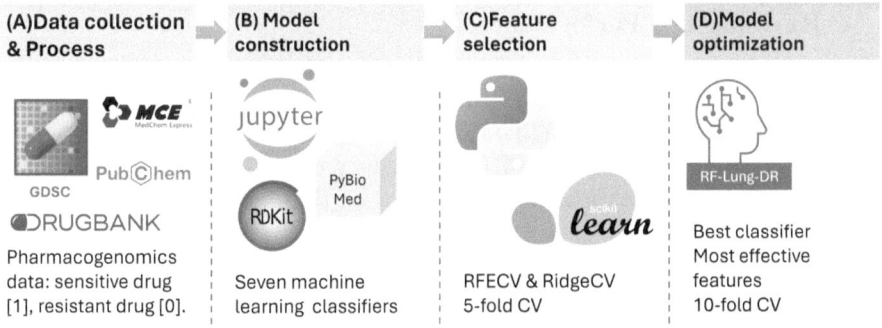

Fig. 1. The framework of constructing RF-Lung-DR.

optimization: The optimal ML classifier is selected, and the model is refined with the most effective features, evaluated by a ten-fold CV. Our optimized RF-based model named RF-Lung-DR, stands for RF-based DRP in lung cancer cell lines.

3.1   Random Forest as the Most Effective Algorithm for DRP in Lung Cancer Cell Lines

We evaluated seven ML classifiers on the LUAD-DR and LUSC-DR datasets. We determined that RF was the most effective classifier, achieving accuracies of 78% for LUAD-DR and 80% for LUSC-DR. Notably, RF demonstrated high sensitivity in predicting drug responses: 81% for LUSC-DR and 77% for LUAD-DR. AdaBoost ranked second, with accuracies of 78% for both datasets. Other models, including SVM, DT, and KNN, showed accuracies around 70%, while Naïve Bayes and LR achieved approximately 60% (Fig. 2). Thus, we selected the RF algorithm to further optimize the DRP model for lung cancer cell lines.

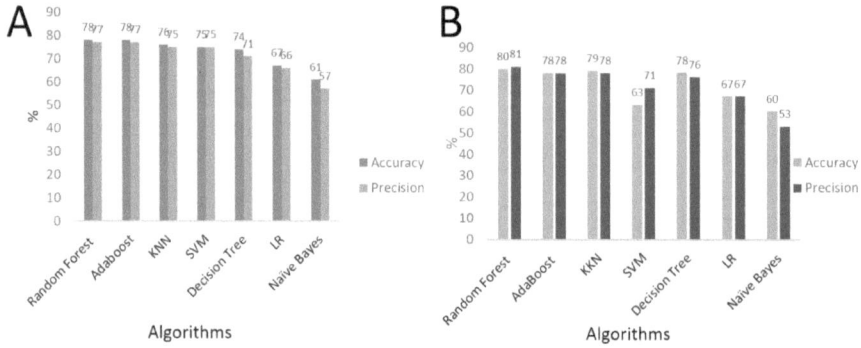

Fig. 2. Different machine learning's performance on (A) LUAD-DR and (B) LUSC-DR dataset.

3.2   Application of Feature Selection Techniques to Enhance RF-Lung-DR Performance

We applied RFECV and RidgeCV to enhance the performance of our RF-based model. The application of these feature selection techniques improved accuracy from 79% to 81% for the LUSC-DR dataset. Notably, RFECV was particularly effective, boosting accuracy to 80% by removing redundant or weak features and retaining only those with a significant impact on training error (Table 1).

3.3   Effectiveness of Drug SMILES Features on DRP Model Performance

We evaluated the impact of biological markers and drug SMILES features on the DRP model's performance by constructing three RF-based models: one using

Table 1. Feature selection techniques to optimize RF-based DRP model.

Dataset	Features selection	Feature number	Accuracy	Precision	Recall	F1-score
LUAD-DR	No	149	0.78	0.78	0.72	0.75
	RFECV	100	0.78	0.77	0.74	0.75
	RidgeCV	105	0.78	0.77	0.72	0.75
LUSC-DR	No	134	0.81	0.80	0.75	0.77
	RFECV	111	0.80	0.78	0.75	0.76
	RidgeCV	75	0.79	0.76	0.76	0.76

only biological features, one with drug features, and a third combining both. The model relying solely on drug SMILES features performed comparably to the combined features model, achieving accuracies of 78% for LUAD-DR and 80% for LUSC-DR. In contrast, models using only biological markers significantly underperformed, highlighting the importance of drug SMILES features in DRP model construction (Table 2).

Table 2. RF-based model's performance on different feature groups selected by the RFECV technique. Models are evaluated by 10-fold cross validation.

Dataset	Feature type	RFECV feature number	Accuracy	Precision	Recall	F1-score
LUAD-DR	Biology	55	0.60	0.58	0.41	0.48
	Drug SMILES	45	0.78	0.79	0.70	0.74
	Combination	100	0.78	0.77	0.74	0.75
LUSC-DR	Biology	60	0.55	0.42	0.12	0.18
	Drug SMILES	51	0.80	0.79	0.73	0.76
	Combination	111	0.80	0.78	0.75	0.76

3.4 Identifying Key Biological Features in the DRP Model for Clinical Applications in Lung Cancer Cell Lines

Using RidgeCV feature selection, we identified critical biological features contributing to DRPs in lung cancer cell lines. For LUAD, significant features included mutations in ASXL2, CTNNB1, MGA, PTEN, and STIP1 genes, along with a gain in copy number alteration (CNA) driven by the KRAS gene. For LUSC, effective features were ARID1A, MLL2, PIK3CA, and ZNRF3 mutations, along with specific CNAs driven by the CDKN2A and TERT genes. These findings underscore the clinical relevance of the identified markers and suggest that the RF-Lung-DR model could provide valuable insights for personalized medicine in lung cancer treatment.

4 Discussions

4.1 Clinical Implications of RF-Lung-DR for Enhancing Personalized Medicine in Lung Cancer Treatment

The advent of precision medicine has significantly transformed therapeutic strategies across various diseases, with cancer treatment at the forefront [20]. The RF-Lung-DR model, developed through our study, represents a significant advancement in personalized lung cancer therapy in two critical aspects: enhancing personalized treatment plans and impacting clinical decision-making.

Personalized medicine aims to tailor therapeutic strategies to individual characteristics, particularly genetic profiles, which vary significantly among lung cancer patients [20]. By integrating biological markers and drug SMILES features, the RF-Lung-DR model predicts the response of lung cancer cell lines to different compounds. This method provides a nuanced understanding of how specific genetic and molecular profiles in lung cancer correlate with drug efficacy. Utilizing such a predictive model allows oncologists to move beyond the conventional trial-and-error approach-which is time-consuming, costly, and often unpredictable [28]-enabling them to select treatments that are more likely to succeed based on the genetic makeup of a patient's tumor. In cases of lung cancer, understanding the sensitivities in different subtypes, such as LUAD or LUSC, can guide clinicians in choosing targeted therapies that are more likely to yield positive outcomes, thereby personalizing treatment plans to a degree previously unattainable.

Furthermore, the RF-Lung-DR model substantially influences clinical decision-making by providing data-driven insights that support more informed and accurate treatment choices. With high prediction accuracies reported (80% for LUSC and 78% for LUAD), the model serves as a powerful tool in clinical settings, where decisions about drug prescriptions must balance efficacy with the potential for adverse effects. Additionally, the model aids in the identification of non-responders to certain treatments, enabling a timely shift to alternative therapies and sparing patients the ordeal of ineffective treatment courses.

4.2 Future Directions of DRP Model to Aid Cancer Treatment

To further aid cancer treatment, future directions for DRP models may include integrating multi-omics data and expanding pharmacogenetics databases to advance comprehensive models.

The integration of multi-omics data-which combines genomics, transcriptomics, proteomics, epigenomics, and metabolomics-provides a comprehensive view of the disease environment [25]. This approach aims to refine prediction accuracy by understanding the complex biological underpinnings of individual responses to treatment. Pharmacogenetics databases play a pivotal role in the future of DRP, serving as foundational resources for personalizing medical treatments based on genetic information. In addition to the GDSC database used in this study, a variety of currently available pharmacogenomics databases are being developed for DRP model development, including the Pharmacogenomics Knowledgebase (PharmGKB) [10], the Cancer Pharmacogenomics Database (PharmacoDB) [8], and COSMIC (Catalogue of Somatic Mutations in Cancer) [23]. These databases are integral to advancing personalized medicine in oncology. They may help researchers identify potential biomarkers for drug response and resistance, support the design of clinical trials, and guide treatment decisions in clinical practice. As genomics technology progresses and more data becomes available, these resources will continue to expand, enhancing their utility in the fight against cancer.

To develop a comprehensive DRP model for clinical use, deep learning (DL), a subset of ML, is a potential approach due to its ability to learn complex patterns and relationships within large datasets. DL models can manage and analyze high-dimensional data effectively, which is common in pharmacogenomics and multi-omics datasets [15, 26]. This capability is crucial for integrating various data types to predict drug responses accurately. Convolutional neural networks (CNNs) and recurrent neural networks (RNNs), which can automatically extract relevant features from raw data, may be particularly effective [27]. In addition, our results suggest that drug SMILES contribute significantly to model performance. Therefore, in the future, focusing on developing DL models that emphasize drug SMILES strings may enhance the model's predictive power.

4.3   Limitations of This Study

Despite our efforts, some limitations remain in this study. Firstly, we collected data from only one database, GDSC2. As a result, it may not address all issues related to different data types and data combinations. In the future, considering pharmacogenomics data from different sources in a single study may provide a comprehensive dataset for model development. Secondly, we focused exclusively on lung cancer, so problems related to other cancers were not solved. Further studies using pan-cancer data are necessary for clinical use.

5   Conclusion

In this study, we developed the RF-Lung-DR model, an innovative DRP model that integrates biological markers and drug SMILES features to accurately predict drug responses in lung cancer cell lines. Our application of RF algorithm, augmented by a comprehensive feature selection process, has proven highly effective, achieving prediction accuracies of 80% in LUSC and 78% in LUAD cell lines. The novelty of this research is that it is the first study for lung cancer to combine genomics and drug SMILES features to construct the model. Furthermore, our results highlight the effectiveness of drug SMILES in enhancing the model's prediction capacity. The methodologies employed here could be adapted to other types of cancer, potentially broadening to pan-cancer. This work, by improving the accuracy of DRP for drug scanning purposes, facilitates more effective and personalized treatment for lung cancer, potentially alleviating the disease's burden and enhancing patient outcomes.

Acknowledgments. This work was financially supported of the Higher Education Sprout Project by the Ministry of Education (MOE) in Taiwan [grant number DP2-TMU-113-A-08].

Disclosure of Interests. The authors have no competing interests to declare that are relevant to the content of this article.

References

1. Adam, S., Feller, A., Rohrmann, S., Arndt, V.: Health-related quality of life among long-term (> 5 years) prostate cancer survivors by primary intervention: a systematic review. Health Qual. Life Outcomes **16**, 1–14 (2018)
2. Ahmad, A.: Breast cancer statistics: recent trends. In: Breast Cancer Metastasis and Drug Resistance: Challenges and Progress, pp. 1–7 (2019)
3. Anderson, E., Veith, G.D., Weininger, D.: SMILES, a line notation and computerized interpreter for chemical structures. US Environmental Protection Agency, Environmental Research Laboratory (1987)
4. Bento, A.P., et al.: An open source chemical structure curation pipeline using RDKit. J. Cheminform. **12**, 1–16 (2020)
5. Chen, J., et al.: Deep transfer learning of cancer drug responses by integrating bulk and single-cell RNA-seq data. Nat. Commun. **13**(1), 6494 (2022)
6. Dong, J., et al.: PyBioMed: a Python library for various molecular representations of chemicals, proteins and DNAs and their interactions. J. Cheminform. **10**, 1–11 (2018)
7. Duffy, D.J.: Problems, challenges and promises: perspectives on precision medicine. Brief. Bioinform. **17**(3), 494–504 (2016)
8. Feizi, N., et al.: PharmacoDB 2.0: improving scalability and transparency of in vitro pharmacogenomics analysis. Nucl. Acids Res. **50**(D1), D1348–D1357 (2022)
9. Feng, F., Shen, B., Mou, X., Li, Y., Li, H.: Large-scale pharmacogenomic studies and drug response prediction for personalized cancer medicine. J. Genet. Genomics **48**(7), 540–551 (2021)
10. Gong, L., Whirl-Carrillo, M., Klein, T.E.: PharmGKB, an integrated resource of pharmacogenomic knowledge. Current Protocols **1**(8), e226 (2021)
11. Han, X., Xie, R., Li, X., Li, J.: SmileGNN: drug-drug interaction prediction based on the smiles and graph neural network. Life **12**(2), 319 (2022)
12. He, X., Liu, X., Zuo, F., Shi, H., Jing, J.: Artificial intelligence-based multi-omics analysis fuels cancer precision medicine. In: Seminars in Cancer Biology, vol. 88, pp. 187–200. Elsevier (2023)
13. Kha, Q.H., Le, V.H., Hung, T.N.K., Nguyen, N.T.K., Le, N.Q.K.: Development and validation of an explainable machine learning-based prediction model for drug-food interactions from chemical structures. Sensors **23**(8), 3962 (2023)
14. Kim, S., et al.: PubChem 2019 update: improved access to chemical data. Nucleic Acids Res. **47**(D1), D1102–D1109 (2019)
15. Le, N.Q.K.: Predicting emerging drug interactions using GNNs. Nat. Comput. Sci. **3**(12), 1007–1008 (2023)
16. Li, B., et al.: A novel drug repurposing approach for non-small cell lung cancer using deep learning. PLoS ONE **15**(6), e0233112 (2020)
17. Misra, P., Yadav, A.S.: Improving the classification accuracy using recursive feature elimination with cross-validation. Int. J. Emerg. Technol. **11**(3), 659–665 (2020)

18. Qureshi, R., et al.: Machine learning based personalized drug response prediction for lung cancer patients. Sci. Rep. **12**(1), 18935 (2022)
19. Schabath, M.B., Cote, M.L.: Cancer progress and priorities: lung cancer. Cancer Epidemiol. Biomarkers Prev. **28**(10), 1563–1579 (2019)
20. Shin, S.H., Bode, A.M., Dong, Z.: Precision medicine: the foundation of future cancer therapeutics. NPJ Precis. Oncol. **1**(1), 12 (2017)
21. Siegel, R.L., Giaquinto, A.N., Jemal, A.: Cancer statistics, 2024. CA Cancer J. Clin. **74**(1), 12–49 (2024)
22. Siegel, R.L., Miller, K.D., Wagle, N.S., Jemal, A., et al.: Cancer statistics, 2023. CA Cancer J. Clin. **73**(1), 17–48 (2023)
23. Sondka, Z., et al.: Cosmic: a curated database of somatic variants and clinical data for cancer. Nucl. Acids Res. **52**(D1), D1210–D1217 (2024)
24. Sotudian, S., Paschalidis, I.C.: Machine learning for pharmacogenomics and personalized medicine: a ranking model for drug sensitivity prediction. IEEE/ACM Trans. Comput. Biol. Bioinf. **19**(4), 2324–2333 (2021)
25. Subramanian, I., Verma, S., Kumar, S., Jere, A., Anamika, K.: Multi-omics data integration, interpretation, and its application. Bioinform. Biol. Insights **14**, 1177932219899051 (2020)
26. Tran, T.O., Vo, T.H., Le, N.Q.K.: Omics-based deep learning approaches for lung cancer decision-making and therapeutics development. Brief. Funct. Genom. **23**, 181–192 (2023). elad031
27. Vo, T.H., Nguyen, N.T.K., Le, N.Q.K.: Improved prediction of drug-drug interactions using ensemble deep neural networks. Med. Drug Discov. **17**, 100149 (2023)
28. Wang, W., Ye, Z., Gao, H., Ouyang, D.: Computational pharmaceutics-a new paradigm of drug delivery. J. Control. Release **338**, 119–136 (2021)
29. Wessolly, M., Schmid, K.W., et al.: Digital gene expression analysis of NSCLC-patients reveals strong immune pressure, resulting in an immune escape under immunotherapy. BMC Cancer **22**(1), 46 (2022)
30. Wishart, D.S., et al.: DrugBank 5.0: a major update to the drugbank database for 2018. Nucl. Acids Res. **46**(D1), D1074–D1082 (2018)
31. Yang, W., et al.: Genomics of drug sensitivity in cancer (GDSC): a resource for therapeutic biomarker discovery in cancer cells. Nucleic Acids Res. **41**(D1), D955–D961 (2012)
32. Yu, L., Xu, Z., Cheng, M., Lin, W., Qiu, W., Xiao, X.: MSEDDI: multi-scale embedding for predicting drug-drug interaction events. Int. J. Mol. Sci. **24**(5), 4500 (2023)

Author Index

H. Chen et al. (Eds.): TAI4H 2024, LNCS 14812, pp. 169–170, 2024.
https://doi.org/10.1007/978-3-031-67751-9